Studies in Computational Intelligence

Volume 786

Series editor

Janusz Kacprzyk, Polish Academy of Sciences, Warsaw, Poland
e-mail: kacprzyk@ibspan.waw.pl

The series "Studies in Computational Intelligence" (SCI) publishes new developments and advances in the various areas of computational intelligence—quickly and with a high quality. The intent is to cover the theory, applications, and design methods of computational intelligence, as embedded in the fields of engineering, computer science, physics and life sciences, as well as the methodologies behind them. The series contains monographs, lecture notes and edited volumes in computational intelligence spanning the areas of neural networks, connectionist systems, genetic algorithms, evolutionary computation, artificial intelligence, cellular automata, self-organizing systems, soft computing, fuzzy systems, and hybrid intelligent systems. Of particular value to both the contributors and the readership are the short publication timeframe and the world-wide distribution, which enable both wide and rapid dissemination of research output.

More information about this series at http://www.springer.com/series/7092

Roger Lee
Editor

Big Data, Cloud Computing, Data Science & Engineering

 Springer

Editor
Roger Lee
Software Engineering and Information
  Technology Institute
Central Michigan University
Mount Pleasant, MI, USA

ISSN 1860-949X   ISSN 1860-9503   (electronic)
Studies in Computational Intelligence
ISBN 978-3-030-07254-4   ISBN 978-3-319-96803-2   (eBook)
https://doi.org/10.1007/978-3-319-96803-2

This Springer imprint is published by the registered company Springer Nature Switzerland AG
The registered company address is: Gewerbestrasse 11, 6330 Cham, Switzerland

Foreword

The purpose of the 3rd IEEE/ACIS International Conference on Big Data, Cloud Computing, Data Science & Engineering (BCD), held on July 10–12, 2018 in Yonago, Japan, was to together researchers, scientists, engineers, industry practitioners, and students to discuss, encourage, and exchange new ideas, research results, and experiences on all aspects of Applied Computers and Information Technology, and to discuss the practical challenges encountered along the way and the solutions adopted to solve them. The conference organizers have selected the best 13 papers from those papers accepted for presentation at the conference in order to publish them in this volume. The papers were chosen based on review scores submitted by members of the program committee and underwent further rigorous rounds of review.

In Chapter "Designing a Method of Data Transfer Using Dual Message Queue Brokers in an IoT Environment", Hee-Yong Kangs, Ji-na Lee, Yoonkyu Kang, Jong-Bae Kim, Hyung-Woo Park, Myung-Jin Bae propose a Bluetooth Low Energy (BLE) plate and Pedestrian Dead Reckoning (PDR) combined algorithm that provides wide range of accuracy and can be applied to indoor positioning for large-scale space. This study resulted in a positioning error within 2.2 m in real environment which is applicable to indoor navigation system for the very large spaces such as airports and arenas.

In Chapter "Bluetooth Low Energy Plate and PDR Hybrid for Indoor Navigation", Sanhae Kim, Hongjae Lee, Kyeong-Seok Han, Jong-Bae Kim propose an application method developed to service from the Activation of Cloud-based Manufacturing Supply Management System (SCM) module for designing SaaS level to the cloud system, which supports SCM tasks such as procurement, purchase, logistics, and standard information as to industrial.

In Chapter "A Study on the Common Collaboration Platform Activation of Cloud-based Manufacturing Supply Management System (SCM)", Haeng-Kon Kim proposes a system for the solution to manage different format models by providing a framework generator model that aims to analyze the top-down framework of the decision problem and the application of the model that aims to integrate the bottom-up model function for a Service Management and Model-Driven Management system.

In Chapter "Service Management and Model Driven Management", Haeng-Kon Kim analyzes the domain modeling support tool that retrieves objects from the candidate domain model to create frameworks from domain descriptions in a typical text format.

In Chapter "Measuring the Effectiveness of E-Wallet in Malaysia", Faisal Nizam, Ha Jin Hwang, and Naser Valaei present a study that aims to discover the important factors influencing consumers' purchase decision using e-wallet. The result of this study indicated that convenience, security, and cost saving were proved to make significant influences on consumers' purchase decision using e-wallet.

In Chapter "Designing of Domain Modeling for Mobile Applications Development", Songai Xuan, Kim DoHyeun build a connection between IoT and cloud, and it is very useful and supports intelligent services based on huge context data. This paper presents the comparison analysis of IoT services based on Clouds for huge context acquisition in large-scale IoT networks.

In Chapter "Performance Analysis of IoT Services Based on Clouds for Context Data Acquisition", Jihyun Lee and Sunmyung Hwang propose a XX-MM-path-based integration testing method, which extends the MM-path-based testing method, and show how test coverage can be handled at both testing levels of domain and application testing for a Path-based Integration Testing of a Software Product Line.

In Chapter "Path-Based Integration Testing of a Software Product Line", Leegeun Ha, Sungwon Kang, Jihyun Lee, and Younghun Han propose a method that automatically generates GUI test inputs under all possible user configurations. Since testing all possible user configurations is infeasible for nontrivial systems, the method is designed such that the user can sample user configurations.

In Chapter "Automatic Generation of GUI Test Inputs Using User Configurations", Jeong Ah Kim analyzes the processes applied to existing R&D projects and standardizes the analytical results into a RD concept, they defined a framework based on SPEM, which further defines method class, method component, and process component as framework components.

In Chapter "Execution Environment for Process defined in EPF", Won-Jung Jang, Soo-Sang Kim, Sung-Won Jung, and Gwang-Yong Gim deduct and suggest factors affecting the intention of big data introduction from the Smart Factory Perspective.

In Chapter "A Study on the Factors Affecting Intention to Introduce Big Data from Smart Factory Perspective", JoongBum Seo, Yong-Won Cho, Kyung-Jin Jung, and Gwang-Yong Gim present a study that focuses on the characteristics of Human Resource cloud service and the effects of the intention to use its technology in an empirical manner. The technology's aspects are organized by researching Human Resource Information system and cloud service and Unified Theory of Acceptance and Use of Technology and as guidelines, preceding studies were used to create the research model and propose the hypothesis.

In Chapter "A Study on Factors Affecting the Intension to Use Human Resource Cloud Service" Seok-Tai Chun, Jihyun Lee, Cheol-Jung Yoo analyze the effectiveness of de-normalization cost and processing time in the very large database based on the case of establishing database for business-to-business service of large retailers. As a result, the de-normalized database had 15% faster processing time at a cost of 0.2% of the normalized.

In Chapter "Comparative Analysis of Cost and Elapsed Time of Normalization and De-normalization in the Very Large Database" Hyun-Seong Lee, Seoung-Hyeon Lee, Jae-Gwang Lee, and Jae-Kwang Lee design a method of data transfer using dual message queue brokers in an IoT environment. Message queue collects the data processing performed by the various services in one place and distributes the work to necessary services by placing a message broker. AMQP is an open standard protocol for message-oriented middleware.

It is our sincere hope that this volume provides stimulation and inspiration and that it will be used as a foundation for works to come.

Yonago, Japan   Akinori Ihara
July 2018   NAIST

Contents

Contributors

Myung-Jin Bae Department Telecommunications Engineering, Soongsil University, Seoul, Korea, South Korea

Yong-Won Cho Department of Business Administration, Soongsil University, Seoul, South Korea

Seok-Tai Chun Department of Software Engineering, Chonbuk National University, Jeonju, Republic of Korea

Kim DoHyeun Department of Computer Engineering, Jeju National University, Jeju City, Republic of Korea

Gwang-Yong Gim Department of Business Administration, Soongsil University, Seoul, South Korea

Leegeun Ha Korea Advanced Institute of Science and Technology, Daejeon, Republic of Korea

Kyeong-Seok Han Department of IT Policy and Management, Soongsil University Graduate, Seoul, Korea, South Korea

Younghun Han Korea Advanced Institute of Science and Technology, Daejeon, Republic of Korea

Ha Jin Hwang Sunway University, Subang Jaya, Malaysia

Sunmyung Hwang Department of Computer Engineering, Daejeon University, Daejeon, Republic of Korea

Won-Jung Jang Department of Intellectual Property for Startups, Catholic Kwandong University, Gangneung, South Korea

Kyung-Jin Jung Department of Business Administration, Soongsil University, Seoul, South Korea

Sung-Won Jung Department of IT Policy and Management, Soongsil University, Seoul, South Korea

Hee-Yong Kang Department IT Policy and Management, Soongsil University, Seoul, Korea, South Korea

Sungwon Kang Korea Advanced Institute of Science and Technology, Daejeon, Republic of Korea

Yoonkyu Kang Korea Telecom, Seoul, Korea, South Korea

Haeng-Kon Kim School of Information Technology, Daegu Catholic University, Gyeongsan, South Korea

Jeong Ah Kim Computer Education Department, Catholic Kwandong University, Gangnung, Korea

Jong-Bae Kim Graduate School of Software, Soongsil University, Seoul, Korea, South Korea

Sanhae Kim Department of IT Policy and Management, Soongsil University Graduate, Seoul, Korea, South Korea

Soo-Sang Kim Department of IT Policy and Management, Soongsil University, Seoul, South Korea

Hongjae Lee Department of IT Policy and Management, Soongsil University Graduate, Seoul, Korea, South Korea

Hyun-Seong Lee Department of Computer Engineering, Han Nam University, Daejeon, Korea, South Korea

Jae-Gwang Lee Department of Computer Engineering, Han Nam University, Daejeon, Korea, South Korea

Jae-Kwang Lee Department of Computer Engineering, Han Nam University, Daejeon, Korea, South Korea

Ji-na Lee Department IT Policy and Management, Soongsil University, Seoul, Korea, South Korea

Jihyun Lee Department of Software Engineering, Chonbuk National University, Jeonju, Republic of Korea; Department of Software Engineering, Chonbuk National University, Jeonju, Korea

Seoung-Hyeon Lee Information Security Research Division, ETRI, Daejeon, Korea, South Korea

Faisal Nizam Sunway University, Subang Jaya, Malaysia

Hyung-Woo Park Department Telecommunications Engineering, Soongsil University, Seoul, Korea, South Korea

JoongBum Seo Department of Business Administration, Soongsil University, Seoul, South Korea

Naser Valaei Sunway University, Subang Jaya, Malaysia

Songai Xuan Department of Computer Engineering, Jeju National University, Jeju City, Republic of Korea

Cheol-Jung Yoo Department of Software Engineering, Chonbuk National University, Jeonju, Republic of Korea

Designing a Method of Data Transfer Using Dual Message Queue Brokers in an IoT Environment

Hyun-Seong Lee, Seoung-Hyeon Lee, Jae-Gwang Lee and Jae-Kwang Lee

Abstract As the number of IoT devices rapidly increases, research is actively conducted to manage data transmitted from a large number of devices. Various services such as monitoring service requesting IoT sensor data and real-time analysis processing service are increasing. However, each device is connected to different services, making expansion difficult. In order to solve this problem, Message Queue collects the data processing performed by the various services in one place and distributes the work to necessary services by placing a message broker. AMQP is an open standard protocol for message-oricnted middleware. It is defined to enable message exchange between different processes or programs. Recently, various services are being provided not only in servers but also in gateways, and message transmission processing is required. In this paper, we propose a method for stable and flexible data delivery by deploying Broker supporting AMQP in gateways and servers in IoT environments where various devices exist.

Keywords Message queue · RabbitMQ · AMQP · Smart gateway

1 Introduction

With the proliferation of IoT connecting things to the Internet, there is a growing problem of connecting many IoT devices. there is a growing problem of connecting

H.-S. Lee · J.-G. Lee · J.-K. Lee (✉)
Department of Computer Engineering, Han Nam University, Daejeon, Korea, South Korea
e-mail: jklee@hnu.kr

H.-S. Lee
e-mail: hslee@netwk.hannam.ac.kr

J.-G. Lee
e-mail: jglee@netwk.hnu.kr

S.-H. Lee
Information Security Research Division, ETRI, Daejeon, Korea, South Korea
e-mail: duribun2@gmail.com

© Springer International Publishing AG, part of Springer Nature 2019
R. Lee (ed.), *Big Data, Cloud Computing, Data Science & Engineering*, Studies in Computational Intelligence 786, https://doi.org/10.1007/978-3-319-96803-2_1

Fig. 1 Existing IoT device
data delivery system

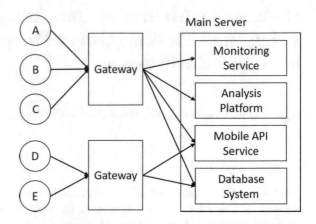

many IoT devices. According to Cisco, IoT connections are expected to increase from 5.8 billion in 2016 to 13.7 billion in 2021 [1]. Many connections to these IoT devices affect not only the server but also the gateway to which the device is connected [2]. Large data traffic and numerous device connections have made it necessary to improve processing performance and increase throughput of gateways, thereby enabling various services such as web server, monitoring service, and analytical processing to be provided within the gateway [3].

On the other hand, in the case of the existing centralized server system, as shown in Fig. 1a lot of device data is transmitted to various services of the server through the gateway. Connection and data processing of various devices are distributed to each application, This is difficult [4]. The concept of Message Oriented Middleware (MOM) has been introduced to deliver data from different systems to multiple services more efficiently [5]. MOM mediates messaging delivery between the different network nodes. By placing the MOM system on the server, it is possible to deliver data efficiently by sending and receiving data asynchronously to various services [6].

However, in the case of a system provided through such a centralized server system, there is a problem that the processing speed is greatly influenced by the server performance and the service latency is delayed [7]. In addition, if the network is disconnected, there is a problem that the service cannot be provided, and the service continuity is lost because the continuously generated IoT device data cannot be transmitted. Recently, as the problem of centralized server and the role of gateway have increased, concepts such as Edge Computing or Fog Computing have emerged [8].

Edge Computing is a concept that solves the problems that arise when only a centralized server is running, and shares its role according to its location and resources on the network [9]. Especially, cloud service is very sensitive to communication speed, stability and security problem because all information is gathered in one place [7]. Edge Computing focuses on gateways connecting and mediating IoT devices. And the gateway can provide the services that were performed in the existing server. The heterogeneous device data that is connected to the gateway must be forwarded to the destination server while delivering the data to the various services within the gateway [9, 10].

In this paper, we propose a method to manage IoT device and data more efficiently in various services existing in gateway and existing server by placing Message Queue Broker in gateway. The rest of the paper is organized as follows: Sect. 2 discusses MOM and AMQP and related research. In Sect. 3, we describe the design of the proposed method and explain it in detail. summarizes this article in Sect. 4.

2   Related Works

2.1   Message Oriented Middleware

Message oriented middleware (MOM) is a system that transports messages between two or more different clients on a network so that they can be routed and delivered [5]. MOM's message communication consists of a simple operation such as sending or receiving a message, and is generally used in an c-mail system, a chat system, and the like. The client calls the API through the MOM system and sends a message to the registered object. The client that sent the message does not need to be involved in the message transfer anymore and can perform other tasks. In addition, even if a network failure occurs, the received message can be processed. If you are using MOM, you can add management interfaces to monitor and extend performance. Therefore, the client can escape problems related to message transmission (Fig. 2).

AMQP is proposed to complement the weaknesses of the Message Queue system among existing Message Oriented Middleware and to exchange messages between various different systems in an efficient manner [11]. Most of the existing commercialized Message Queue products are platform-dependent products, and there is a problem that message bridges for message format conversion must be used to exchange different types of messages, or the system itself must be unified. AMQP 1.0 is an international standard approved by ISO and IEC, developed by over 20 major

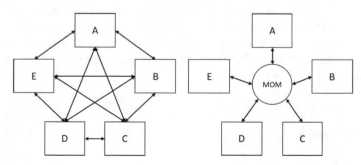

Fig. 2 Message oriented middleware concepts

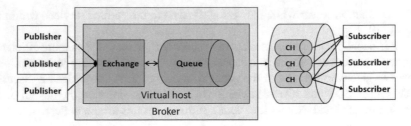

Fig. 3 AMQP's overall flow diagram

companies including Microsoft, Axway and Huawei. As an open standard, there
are advantages such as interoperability, various libraries and tools, and antiquated
protection.

It is a binary protocol with the latest technology and has features of Multi-Channel,
Negotiated, Asynchronous, Secure and Efficient. The AMQP must meet the following conditions in order to be neutral to various systems.

- All brokers should behave in the same way
- All clients should behave in the same way
- Standardization of commands transmitted over the network
- Programming language should be neutral

The AMQP routing model is constructed as shown in Fig. 3 Exchange routes
messages received from Publisher to a queue or other Exchange. Message Queue is
a commonly used queue. It temporarily stores the message in memory or disk and
sends it to the consumer. The queue is bound directly to Exchange through a Binding
that specifies the message type. The same queue can be bound to multiple Exchange,
and multiple queues can be bound to a single Exchange.

Exchange distributes messages received from Publisher to the appropriate queue
or to another Exchange. Each Queue or Exchange is bound to one or more Exchange
through the Binding function. We defined a routing algorithm to match the message
with the binding, which is called the Exchange Type. Because Exchange Type determines where and how to deliver the message, it is actually a routing table that determines which messages are queued to which queue. The actual connection between
Queue and Exchange consists of Binding. In AMQP, one of the three types of routing
algorithms based on the routing key based on the routing key, Topic Exchange, and
Fan-out Exchange and one of Header Exchange based on the key-value header must
be defined.

2.2   Smart Gateway

Aazam and Huh [12] Paper presents limitations of cloud computing in the IoT era
and suggests methods for more useful and efficient service provisioning and resource

utilization. We discuss the need to pre-process and service the data generated by IoT devices within the smart gateway. The limitations of cloud computing and the role of smart gateways have been summarized, but no specific solution has been suggested.

In [6] paper, we propose a messaging bus-based middleware for mass exchange of data in a smart grid environment. Analyze existing messaging buses and talk about how to apply them. The structure and characteristics of Extensible Messaging and Presence Protocol (XMPP), AMQP, and DDS (Data Distribution Service) are described and their applications are introduced. In conclusion, DDS is suitable for global-scale real-time services, but it is not suitable for centralized server systems that are an existing infrastructure because they must be distributed directly to devices and are device-to-device. The AMQP is focused on stable interoperability and interoperability, which is relatively slow, but has the advantage of being easy to apply and reliable on a variety of platforms. However, AMQP has a problem in that it is applied when communication between a plurality of brokers is made because communication between brokers is not defined.

In [13] paper, we analyze the requirements for publishing/subscribing to IoT platform and analyze the requirements of popular protocols such as RabbitMQ (AMQP), Mosquitto (MQTT), ejabberd (XMPP) And ZeroMQ is performance tested and compared. XMPP is an open, extensible protocol, but there is a problem with parsing all 'Stanza' in order to make routing decisions on the server side. In addition, XMPP performed worse than other protocols. The MQTT is designed to transmit small data such as IoT sensors quickly, making it a de facto standard. Experimental results show relatively high performance, but since MQTT is a pure transport protocol, the structure and performance of the broker are important and there is a problem that it is directly connected to the client. ZeroMQ achieved very high throughput while maintaining low latency without significantly affecting the load. There is also the advantage that a broker is optional and a distributed system is possible. However, ZeroMQ is not built with brokers in mind, and has low expressiveness that is only supported for prefix matching. AMQP, on the other hand, is a MOM that provides all the components necessary to create a messaging broker for IoT. Message throughput and average delay performance are less than ZeroMQ and MQTT in the presence of a large number of messages, but they are advantageous for delivering data to services in a centralized server because there are well-defined brokers that are advantageous for interoperability across various platforms.

In this paper, we propose to deploy MOM as a method to efficiently transfer data of various IoT devices to services in smart gateways and servers. The AMQP is suitable for operation in addition to existing systems using a centralized server, and can selectively receive device data with various rules in various services. It can also pass data in many formats. Therefore, in the proposed method, MOM is designed using AMQP.

3   Data Transfer Using Dual Message Queues

In the existing system, as shown in the Fig. 4, there is a message queue only in the
server, and the data is transmitted to various services through the gateway. The role
of the gateway here is simply to pass data to the destination. In this paper, AMQP
is deployed in smart gateways and servers that provide various IoT services, and we
propose a method to transfer IoT device data more stably and efficiently.

AMQP is a well-defined MOM protocol and there are various platforms that
implement it. Among them, RabbitMQ is a message broker implemented based on
AMQP. It is easy to configure the cluster and Web Management UI is provided. In
addition, plugins such as MQTT Convert and STOMP are also provided, making it
easy to connect to various platforms with excellent scalability. The message broker
uses RabbitMQ as the message delivery method proposed in this paper [14].

A device belonging to a gateway can send data according to the standard protocol,
MQTT, or send it according to the format supported by the message queue broker.
When a message arrives at the gateway's message queue broker, messages are queued
up sequentially in the queue. This queue can be stored in memory or on disk. There
are several services in the gateway, and the device data required for each service is
different. Each service in the gateway subscribes to the device data required by the
message queue broker according to the AMQP standard.

3.1   IoT Device to Gateway Message Broker

The IoT device data transmitted to the gateway includes an ID value for identifying the
device, a Body value containing the contents, and an Exchange type. The gateway's
AMQP broker consists of one or more queues and exchanges. The transferred data
is passed to the specified queue and delivered to one or more consumers according
to the Exchange type of the queue. Messages delivered to the Queue can be stored
in two ways. Persistent messages are written to disk as soon as they arrive at the

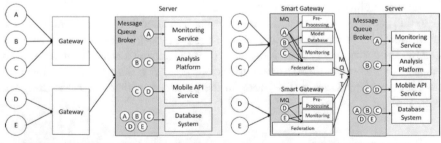

Existing centralized servers and generic gateways   Servers and Smart Gateways with Message Queue

Fig. 4 IoT data delivery method of existing server and proposed smart gateway and server

File descriptors ?	Socket descriptors ?	Erlang processes	Memory	Disk space
27	2	380	194MB	326GB
1048576 available	943626 available	1048576 available	6.2GB high watermark	48MB low watermark

Fig. 5 Monitoring broker resources

queue, while temporary messages are written to disk when messages are removed from memory in low memory situations. Persistent messages are kept in memory as long as possible and are removed from memory in the event of insufficient memory to prevent performance degradation due to the slow I/O rate of the disk. The gateway is configured to use only temporary messages and only as much of the resources as memory is small and there is little need to keep the data long. In RabbitMQ Management, the remaining amount of memory and file descriptor can be checked as shown in Fig. 5 to easily check the problem situation.

The AMQP network configuration between the IoT device and the gateway is configured in the same way except for the OS kernel parameters and DNS. In order for the AMQP Broker to connect to the device, it must connect to one or more interfaces and list the ports. If the number of connected devices is low, the rate of new connections will not be evenly distributed, but will have negligible impact. However, if there are tens of thousands of connections, the server may not be able to receive incoming connections. Unconnected TCP connections enter a lengthy queue. The length of this queue should be set in consideration of peak load time and surge.

3.2 Gateway Message Broker to Gateway Services

The service in the gateway receives the message arriving asynchronously after having the binding for the queue through the message queue broker. The message generated by the device and delivered to the Queue is determined by Exchange. It defines how Exchange will deliver the message to the Queue in the broker and routes the message according to the defined rules. Exchange Type is a way of determining Exchange rules. Direct delivers the message only to the queue with the routing key specified in the unicast method. Fan-out is like broadcast in that it sends messages to all known queues. topic will only deliver messages to Queues that match the pattern binding type specified. You can specify a string pattern match using multiple words with the '#' character and a word with the '*' character. Finally, the header carries the message according to the key = value matching condition contained in the header. At the bottom of Fig. 6, 'custom.topic_log' Exchange is a rule adding an arbitrary string. The advantage is that you can configure Exchange directly so that you do not have to separate a lot of data from many services.

The service can be configured to receive notification of receipt of the message in case the data is not received. The AMQP Broker uses ack(Acknowledgment) for this purpose. If you do not use ack, it will use ack automatically. If you set manual ack

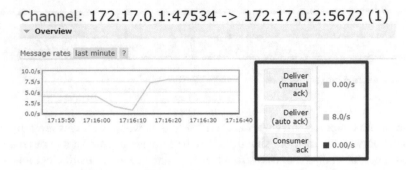

Fig. 6 The acknowledgment status of the message passed on the connection

Queues

▼ All queues (6)

Pagination

Page 1 ▼ of 1 - Filter: [   ] ☐ Regex ?

Overview			Messages			Message rates		
Name	Features	State	Ready	Unacked	Total	incoming	deliver / get	ack
GW_Group1_Queue	D	idle	0	0	0			
Gateway_A	D	running	0	0	0	4.0/s	4.0/s	0.00/s
Gateway_B	D	running	0	0	0	4.0/s	4.0/s	0.00/s
Gateway_C		idle	0	0	0			
Gateway_D	D	idle	0	0	0			
Gateway_E		Idle	0	0	0			

Fig. 7 Status information and monitoring of queues

as the transmission option, you will receive a notification. Figure 6 is to check the status of this ack. If the service does not send an ack, the queued messages are not erased. When the service is restarted again, the service sends the request again and the Queue can deliver the stacked message.

Also, the Broker may terminate with a message arriving at the Queue and not reaching the Service. In this case, you can set the durable property when creating the queue. Figure 7 shows the Queue list and its status and details. Features Queued with D is set to Durable. Durable can be written to disk as soon as the message arrives, so you can keep messages until the network is down or the broker is about to shut down. Messages in the figure represent messages that are not delivered or messages that are waiting to be delivered.

3.3 Gateway Message Broker to Server Message Broker Using Federation

Smart gateways need to transmit large amounts of device data to the cloud or server while being able to service themselves. The server also puts MOM in place to deliver large amounts of data to multiple services. If you apply the basic Queue method to the data passing between two Brokers, messages will be awaited by a large number of queues. Also, the transmitted data is slowed down because the broker goes through the queue twice. Federation Exchange links these broker-to-broker communications efficiently. Federation Exchange is unidirectional, and two brokers are directly connected. It is typically used to forward to Exchange connected to another broker. To connect to the Federation, you must configure Upstream, which is the node that transmits data to the central server. Upstream is created as a virtual host, one for each gateway, so you have to define Exchange internally in the same state as the new Broker.

Figure 8 shows the screen to create a new Upstream. Name is upstream's name, URI is the address of the Broker to connect to, and Expires is the message retention time (if empty, it is stored until the memory is full). The Acknowledgment Mode option determines the degree of message delivery and is divided into on-confirm, on-publish, and no-ack. On-confirm sends a completion message to the Broker sending the message to the Broker receiving the message, which can prevent loss due to network or broker error. On-publish is sent when a Broker that sends a message

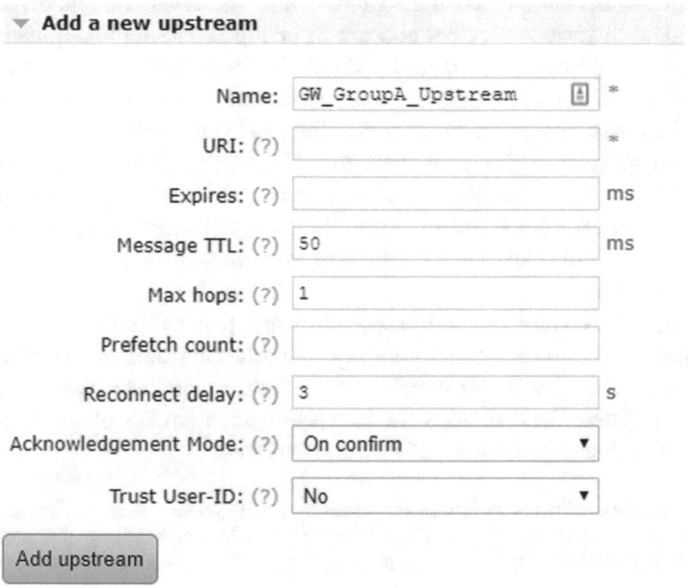

Fig. 8 New upstream creation and generation information

sends it, so it can prevent the loss by the network, but it cannot prevent the loss caused by the broker's error. no-ack is sent without confirmation and there is no processing for loss. Depending on the purpose of the gateway and the characteristics of the data, the upstream can be configured in various ways.

The Shovel Exchange operates at a lower level than the Federation Exchange, which forwards messages from one broker to Exchange from another broker. The Shovel Exchange can be configured to connect all nodes, allowing for a wider variety of delivery methods. In this paper, since unidirectional transmission is performed from the smart gateway to the server, it is configured to enable efficient message delivery between brokers using Federation.

However, the Federation Exchange provided by RabbitMQ can only deliver messages in AMQP format. AMQP has overhead for data transfer due to various functions and options. Communication between the gateway and the server's broker does not use complex functions or options. Therefore, MQTT can be applied to message delivery between brokers to enable faster delivery.

4   Conclusion and Future Work

In this paper, we propose a method that can efficiently and flexibly transfer data by using Message Queue to various services of main server in smart gateways to which many IoT devices are connected. It supports various formats including AMQP, an open standard protocol, so data from different devices can be managed at once. The Message Queue Broker located at the gateway and the server can reliably deliver the necessary data in various services and can be changed and applied quickly without any complicated process even if new services or devices are added. Recently, with the concept of Edge Computing, various services are provided at the gateway. We designed dual message queue brokers to deliver IoT device data to various services in this smart gateway and deliver them to the server quickly. For the direct communication between the brokers, a virtual host is created and connected inside, and the MQTT is used for faster transmission. Reliability can also be increased by using persistent queues and acknowledgments to prevent message loss due to broker errors or network failures.

However, the delivery method proposed in this paper differs depending on the performance and resources of each gateway. Also, the environment in which multiple gateways are connected is not considered. In future work, we plan to design and implement a system that connects the Message Queue Broker of smart gateway to parallel or distributed processing of data and services.

Acknowledgements This research was supported by Basic Science Research Program through the National Research Foundation of Korea(NRF) funded by the Ministry of Education(NRF-2017R1D1A3B03036130). This work was supported by 2018 Hannam University Research Fund.

References

1. Cisco: Cisco visual networking index: global mobile data traffic forecast update, 2016–2021 white paper. https://www.cisco.com/c/en/us/solutions/collateral/service-provider/visual-ne tworking-index-vni/mobile-white-paper-c11-520862.html#EvolvingtowardSmarterMobile. Accessed 10 Apr 2018
2. Zachariah, T., Klugman, N., Campbell, B., Adkins, J., Jackson, N., Dutta, P.: The internet of things has a gateway problem, 27–32 (2015)
3. Simeunović, M., Mihailovic, A., Pejanović-Djurišić, M.: Setting up a multi-purpose internet of things system. In: 2015 23rd Telecommunications Forum Telfor (TELFOR), pp. 273–276 (2015)
4. Rausch, T., Wien, T.: Message-oriented middleware for edge computing applications, 3–4 (2017)
5. Curry, E.: Message-oriented middleware. Middleware for Communications (2004)
6. Albano, M., Ferreira, L.L., Pinho, L.M., Alkhawaja, A.R.: Message-oriented middleware for smart grids. Comput. Stand. Interfaces **38**, 133–143 (2015)
7. Aazam, M., Huh, E.N.: Fog computing and smart gateway based communication for cloud of things. In: 2014 International Conference on Future Internet of Things and Cloud, pp. 464–470 (2014)
8. Perera, C., Qin, Y., Estrella, J. C., Reiff-Marganiec, S., Vasilakos, A.V.: Fog computing for sustainable smart cities: a survey. ACM Comput. Surv. (CSUR), **50**(3) (2017)
9. Bonomi, F., Milito, R., Zhu, J., Addepalli, S.: Fog computing and its role in the internet of things. In: Proceedings of the First Edition of the MCC Workshop on Mobile Cloud Computing, pp. 13–16 (2012)
10. Davis, A., Parikh, J., Weihl, W. E.: Edgecomputing: extending enterprise applications to the edge of the internet. In: Proceedings of the 13th International World Wide Web Conference on Alternate Track Papers & Posters, pp. 180–187 (2004)
11. Home | AMQP: Amqp.org. https://www.amqp.org/. Accessed 12 Apr 2018
12. Aazam, M., Huh, E. N.: Fog computing and smart gateway based communication for cloud of things. In: Future Internet of Things and Cloud (FiCloud), pp. 464–470. IEEE (2014)
13. Happ, D., Karowski, N., Menzel, T., Handziski, V., Wolisz, A.: Meeting IoT platform requirements with open pub/sub solutions. Ann. Telecommun. **72**, 41–52 (2017)
14. RabbitMQ—Messaging that just works. http://www.rabbitmq.com/. Accessed 10 Apr 2018

Bluetooth Low Energy Plate and PDR Hybrid for Indoor Navigation

Hee-Yong Kang, Ji-na Lee, Yoonkyu Kang, Jong-Bae Kim, Hyung-Woo Park and Myung-Jin Bae

Abstract Bluetooth Low Energy(BLE) is becoming the most preferred technology for indoor positioning solutions as it provides a low cost, reasonably accurate, and zero infrastructure solution, and is supported by most mobile devices as well. BLE can maintain relatively stable signal strength and has a higher sample rate, which is important for accurate distance determination. On the other hand, RSSI values are not always accurate due to fluctuating signal strength for many reasons, so a designated process and algorithm are used to determine the actual location of the mobile device. In this paper, we propose a BLE plate and Pedestrian Dead Reckoning (PDR) combined algorithm that provides a wide range of accuracy and can be applied to indoor positioning for large scale space. The system tested in this study kept positioning errors to within 2.2 m in real environment, and is applicable as an indoor navigation system for very large spaces such as airports and arenas.

Keywords BLE plate technique · Plate and PDR hybrid · Indoor positioning

H.-Y. Kang · J. Lee
Department IT Policy and Management, Soongsil University, Seoul, Korea, South Korea
e-mail: hykang07@naver.com

J. Lee
e-mail: ppjina@hanmail.net

Y. Kang
Korea Telecom, Seoul, Korea, South Korea
e-mail: ssme2@naver.com

J.-B. Kim (✉)
Graduate School of Software, Soongsil University, Seoul, Korea, South Korea
e-mail: kjb123@ssu.ac.kr

H.-W. Park · M.-J. Bae
Department Telecommunications Engineering, Soongsil University, Seoul, Korea, South Korea
e-mail: pphw@ssu.ac.kr

M.-J. Bae
e-mail: mjbae@ssu.ac.kr

© Springer International Publishing AG, part of Springer Nature 2019
R. Lee (ed.), *Big Data, Cloud Computing, Data Science & Engineering*, Studies in Computational Intelligence 786, https://doi.org/10.1007/978-3-319-96803-2_2

1   Introduction

Bluetooth Low Energy uses 40 channels on the 2.4 GHz, three of which are advertising channels used for device discovery. The received signal strength (RSS) of these
channels from multiple BLE devices is used to determine relative distance and location. Transmission power for BLE devices can be set from 0 dBm to −80 dBm and
the advertising rate can be configured up to 50 Hz. Typically, transmission power is
set to less than −16 dBm and the advertising rate to less than 10 Hz, for the purpose
of energy conservation [1, 3, 7].

Depending on frequency and signal power, BLE devices can run for years on
a coin cell battery. Reduction of power consumption is achieved by the very short
duration of the messages, which can be either a data message or an advertising
message. Advertising messages, which are broadcast at periodic intervals, are used
for proximity and positioning. BLE can maintain relatively stable signal strength and
has a higher sample rate, which is important for accurate distance determination [3,
4, 7].

Pedestrian dead-reckoning (PDR) has been extensively studied as an effective
approach to obtaining pedestrian locations by estimating the distance traveled via
handheld inertial sensors [4]. Recently, two main types of algorithms for distance
estimation have been described. The first type is based on the successive double-
integral-based length-step measurement of acceleration. The major drawback of this
technique is the error accumulated over the duration of the experiments. This problem can be partially addressed using zero velocity updates [4]. In the second type,
researchers applied a verifiable relationship between vertical acceleration and the
step length to estimate the distance traveled by a moving subject.

In this paper, we propose a plate method that reduces the low accuracy and large
positional deviation problems that are a weakness of the existing BLE based RSSI
proximity method, and an algorithm which improves the accuracy and efficiency of
the indoor positioning system by converging the Pedestrian Dead Reckoning (PDR)
method using IMU data of the mobile device and the Plate method.

2   Overview of Proposed Works

2.1   BLE Plate Technique

Bluetooth beacons are hardware transmitters, a class of Bluetooth Low Energy (BLE)
devices that broadcast their identifier to nearby portable electronic devices. The
technology enables smartphones, tablets and other devices to perform actions when
in close proximity to a beacon.

Bluetooth beacons use Bluetooth low energy proximity sensing to transmit a
universally unique identifier picked up by a compatible app or operating system. The
identifier and several bytes sent with it can be used to determine the device's physical
location, track customers, or trigger a location-based action on the device such as a
check-in on social media or a push notification.

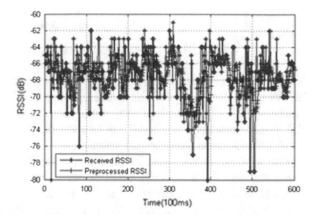

Fig. 1 Pre-processed RSSI values: as can be observed in the figure, the pre-processed RSSI has less spikes in the data curve [1]

In this paper, we propose an algorithm that uses the plate technique to address a weakness of the conventional RSSI positioning method. The plate consists of pre-process of RSSI, the application of Kalam Filter and trilateration. The plate is a circle with a radius of 5 m, which is created when the device is moving. Later, those circles will be guided to calibrate Pedestrian Dead-Reckoning (PDR) path estimation.

First, the RSSI value is preprocessed. The RSSI value of each beacon received in the process of scanning the RSSI signal of the BLE beacon is greatly dispersed by multipath fading. As shown in Fig. 1, the singular value is first removed, the distinguished valuable RSSI is then valued using the moving average method, and eventually the RSSI value is refined to obtain a useful RSSI value. After the RSSI preprocessing process, a scan list is generated that is added to or updated according to whether the preprocessed useful RSSI value is new or not.

A scan list is created by eliminating outliers and calculating the moving average, the mean and standard deviation, and calculating the coordinates and distance of the BLE beacon through RSSI preprocessing to distinguish only useful RSSI values. The average value of the RSSI is calculated and used as the preprocessed value in the next step.

Depending on whether a preprocessed RSSI value is new, it is determined whether the preprocessed RSSI value is filtered by the Kalman filtering technique or only updates the Kalman filter list. Figure 1 shows the results after RSSI preprocessing.

Secondly, a Kalman filtering process is applied. The distance between the BLE beacon and the mobile device is calculated using the pre-processed RSSI value from previous step. For more precise distance calculation, the RSSI value calculated through the pre-processing process is calibrated by applying the actual distance calculation Eq. (1) of the Kalman filtering algorithm, where rssi_p is the preprocessed RSSI value and rssi_cali is 1(one) meter distance based on the calibrated value [2, 5].

$$dist = \begin{cases} 10^{\frac{rssi\_p}{rssi\_cali}}, & rssi\_p > rssi\_cali \\ 0.9 \times 7.71^{\frac{rssi\_p}{rssi\_cali}} & rssi\_cali < rssi\_cali \end{cases} \qquad (1)$$

The Kalman filtering algorithm of Eq. (1) is applied to the pre-processing processed values to more accurately calculate the distance between the mobile device and the BLE beacon.

Thirdly, a Trilateration process is applied. To reduce the deviation between the distance calculated by the Kalman filtering and the actual distance, choose the three BLE beacons closest to a mobile device, calculate the distance from the device to each three beacons, and find the intersection point of three circles derived by the beacons. The intersection point is recognized as the device location.

The distance is calculated using a standard RSSI attenuation model (2) based on the pre-computed distance around each.

$$d_m = 10^{\frac{RSSI-A}{-10.n}} \qquad (2)$$

Figure 2. shows the device moving path as the result of Kalman Filtering, The figure on the left is pre-processing only and the figure on the right is with Kalman Filtering after RSSI pre-processing.

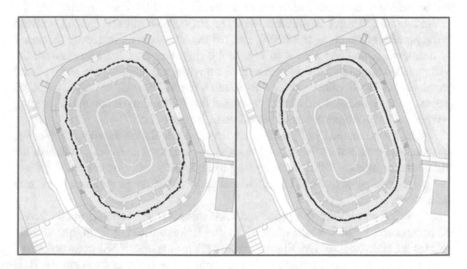

Fig. 2 Device moving path before (left) and after Kalman Filtering (right)

The characteristics of the plate technique are that it minimizes the computation involved in calibration and improves the reliability of the BLE beacon RSSI, while maintaining the immediacy regardless of the positioning data amount. The plate technique uses a beacon scanning process to perform reliable RSSI filtering and derive a discriminative RSSI against the distance, and it repeats the process until the discriminant RSSI value is derived within an average of 1 second.

2.2 Pedestrian Dead-Reckoning

PDR derives an estimation of new position from the first fix points based on step stride, step counting and heading. To measure step counts and estimate stride lengths and headings, IMUs such as accelerometers, gyroscopes, barometers and magnetometer are used, and also, absolute position (starting point) coming from Wi-Fi, BLE is used. Through repetitively updating pedestrian's walking distance and direction, Pedestrian location can be estimated. The basic mechanism of pedestrian dead reckoning is as shown below.

$$\begin{bmatrix} x_k \\ y_k \end{bmatrix} = \begin{bmatrix} x_{k-1} + l_k \sin(\theta_k) \\ y_{k-1} + l_k \cos(\theta_k) \end{bmatrix} \tag{3}$$

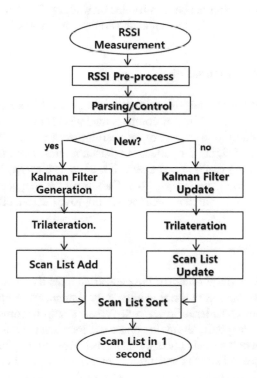

Fig. 3 Scan list generation for plate

where l is the step length, θ is the estimated current heading and (x, y) are the coordinates in the horizontal plane. Index k is an abbreviated form of t(k) denoting the number of a discrete point within a given time [5–7].

2.2.1 Step Length Estimation

A pedestrian's stride length is closely related to height. In general, a normal stride length is estimated as 37% of height but can be as high as 45% during exercise. Accelerometer output can be used for step detection, step count and step length estimation. It is necessary to categorize pedestrian physical activity into walking, exercising, running, ascending/descending stairs, taking an elevator, standing still and irregular movements. Different dynamics in the movements can be recognized using data from the accelerometers integrated into a mobile device which is carried on the waist or held in the hand. Where and how the device is carried by the pedestrian are decisive factors in estimating step count and step length.

The number of steps is calculated to determine the distance travelled, and the stride lengths are estimated based on the frequency of step. As shown in Fig. 2, step length increases linearly with the step frequency. In addition, changes in speed also indicate the increase or decrease of stride. Once an absolute position of the starting point is available, the stride can be recalibrated on walking. The combined calibration technique using IMU is applied to a specific context or a new situation.

2.2.2 Step Heading Estimation

The heading can be determined with azimuth reading a leveled compass with a typical accuracy of 5° [7]. The reliability of compass readings is compromised by magnetic perturbations caused by power lines or iron reinforcement and misalignment of the device (i.e. the line of sight of a pedestrian does not correspond to the direction of walk) [7]. The heading can be estimated using a magnetometer, but those data for direction determination have many errors. Data from the magnetometer should be tuned using a gyroscope, which is necessary to improve the reliability of estimation.

2.2.3 Step Counting

The mobile device is tightly held in the pedestrian's hands when walking straight ahead. Then the orientation will not significantly change, and step detection would be possible without considering drawback. When a step is composed of a swing phase and a heel strike phase, the swing phase occurs when the reference foot moves forward from from behind the contralateral leg [4]. Heel strike phase happens when the heel contacts ground and the waist (the center of gravity) is in its lowest position during the entire step. In the swing phase, a vertical acceleration begins from a minimum valley, then goes to a maximum peak, and finally stops at another minimum

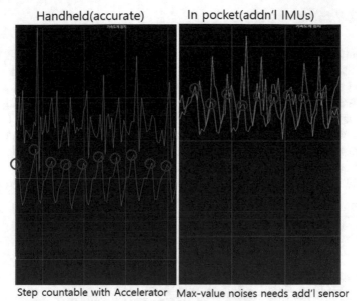

Fig. 4 Step counting deviation on device position: the curve in the figure on the right shows much noise at max values when mobile device is in pocket

valley, which is caused by the next heel strike phase [4]. In other words, the vertical acceleration reaches maximum peak and minimum valley within a step. Figures 2, 3 and 4 illustrates the pattern of the vertical acceleration in a step [3].

Accelerometers measure the number of steps, when a person is walking with a handheld device, the accelerometer can accurately measure the number of steps, but if the device is kept in the pocket, it will generate a lot of noise and thus the number of steps should be measured in combination with other IMUs and a special algorithm as shown in Fig. 4. Where and how the device is carried by the pedestrian is decisive in estimating step count and step length.

3 Proposed BLE Plate and PDR Hybrid for Indoor Positioning

In this paper, we propose a converged technique for indoor positioning that combines BLE Plate and PDR technique for high precision to address the weaknesses of the existing indoor positioning method.

3.1   BLE Plate

BLE Plate technique uses Scan list results from the pre-process RSSI Value, Kalman filtering and trilateration process. A scan list is created by calculating the coordinates and distance of BLE Beacon through an RSSI pre-processing process that distinguishes the useful RSSI values by removing outliers and calculating a mean and standard deviation with moving average. The distance between BLE and the device is calculated using the pre-processed RSSI value, and the Kalman filtering algorithm is applied to the pre-processed values to more accurately calculate the actual distance between the device and BLE.

The following is a part of the program code for RSSI calibration using Kalman Filtering.

```
Public double getBearing() {
  double x, y;
  Double[] latlon = getLatLong();const  MaxYears = 10;
  Double[] delta_latlon = getVelocity();

  latlon[0] = Math.toRadians(latlon[0]);
  latlon[1] = Math.toRadians(latlon[1]);
  delta_latlon[0] = Math.toRadians(delta_latlon[0]);
delta_latlon[1] = Math.toRadians(delta_latlon[1]);

  double lat = latlon[0] - delta_latlon[0];
  y = Math.sin(delta_latlon[1]) - delta_latlon[0];
  x = math.cos(lat1) * Math.sin(latlon[0]) -
Math.sin(lat1) * Math,cos(latlon[0]) *
Math.cos(delta_latlon[1]);
  double bearing = Math.atan2(y,x);

  bearing - Math.toDegrees(bearing);  return bearing;
  return
}
```

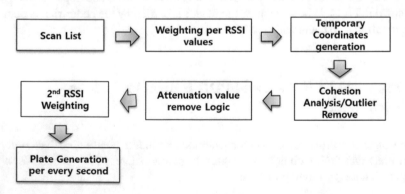

Fig. 5 Plate generation process: To obtain precise plate coordinates, the BLEs already generated in the scan list are weighted to then get temporary coordinates. After cohesion analysis, attenuation value is removed, 2nd RSSI is weighed, and eventually Plate is generated to be used for PDR hybrid.

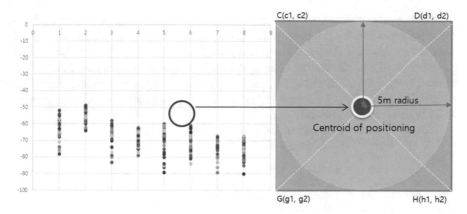

Fig. 6 Generated plate based on optimized RSSI value

Next, in order to reduce the deviation from the actual distance, trilateration is performed, which calculates distance of the three BLE's closest to the mobile device to determine the device location.

For plate generation, the weight logic is applied to the BLE's in the scan list to weigh each RSSI value of each beacon. Weights are assigned to RSSI values of the scanned coordinates by applying weighting algorithm to each interval, and temporary coordinates are generated as an average value of the weighted beacon coordinates. Temporary coordinates consist of latitude, longitude, floor information, and location information (Indoor Map Data, Stored POIs) from the server.

The next step is to analyze extracted beacon coordinates to distinguish beacons that are excessively deviated on the basis of the temporary coordinates (TC) and to remove the beacons from target beacons for positioning, this is cohesion analysis and removal process.

After the above process of cohesion analysis and outlier removal attenuation, useless value removal logic is applied to select and remove beacon(s) out of the scan list for which the RSS value is no longer scanned, and beacon(s) for which the same RSS value is continuously measured for a predetermined time.

The last step is 2nd weighting algorithm which applies to beacons with valid RSSI values, and then RSSI values are sorted to define Plate.

The plate is a positioning plate that includes information on the position of the mobile device, such as latitude, longitude, radius of mobile device, effective beacon and information generation time. This process is generated every second. The algorithm of the positioning plate and generated plate are shown in Figs. 5 and 7, respectively.

The plate is a circle with a radius of 5 m with a centroid determined as a result of repetitive RSSI value filtering experiments (-60 dB).

Therefore, the BLE in the plate is used for positioning determination. That is, C, D, G, and H are the beacons to be subjected to indoor positioning calculation.

When a mobile device moves, new coordinates which are subjected to the BLE calibration process by the plate technique are calculated again. The re-calculated coordinates are used for positioning and used as a final set of Plates to be applied to Pedestrian Dead Reckoning technique to correct the movement path of a mobile device. In applying PDR technique, the movement path is limited within a 5m radius of the plate plate. Part of the program code for RSSI calibration in order to generate BLE Plate is provided below.

```
Public static Pzlocation getNearPoint(PzLocation
  startLoc, Pzlocation enfLoc, PzLocation gpsLoc)
  {
    Double m1 = (endLoc.getLongitude() -
  startLoc.getLongitude());
    Double m2 = (endLoc.getLatitude() -
    startLoc.getLatitude());

    if(m1 == 0.0) m1 = 0.0000001;
    if(m2 == 0.0) m1 = 0.0000001;

    Double m = m1/m2;
    Double n = -1.f/m * 88.607f.110.950f;

    Double lat = (-startLoc.getLongitude() + m *
    startLoc.getLatitude() + gpsLoc.getLongitude(0 - n *
    gpsLoc.getlatitude())/(m-n);
    Double lng = m*lat - m*(startLoc.getLatitude() +
    startLoc.getLongitude();

    Return new PzLocation(lat, Lng);
  }
```

The top figure of Fig. 7 shows that, the reliable beacons C and H close to −60 dB were selected out of BLE beacon C, D, G, and H after evaluating the RSSI value of each beacon to calculate the center of the plate. A BLE scan range of 5 m radius, that is a plate 1, is centered on the calculated coordinates. ◐ is ground truth and actual position, and ● is estimated center of plate.

The 2nd figure from the top explains that Beacons C and G are selected as reliable RSSI values after scanning beacon C(−58dBm), G(−65dBm), H(−78dBm) and D. D is disregarded due to its significant deviation. And the trilateration technique with C, G beacon RSSIs applied to generate the centroid of Plate 2

The 3rd figure explains that outlier 3 is calibrated and moved into the plate. In the 4th figure, becoans A, B, E, and F beacons are all scanned, but RSSI value is −60dBm, −60dBm, −78dBm and 77dBm respectively. Beacon A and B are used for trilateration to find the centroid of plate 4.

The bottom figure illustrates errors of proximity from ground truth. Each point estimation is much closer to the actual location, but additional calibration is still required to be useful for the purpose of indoor navigation.

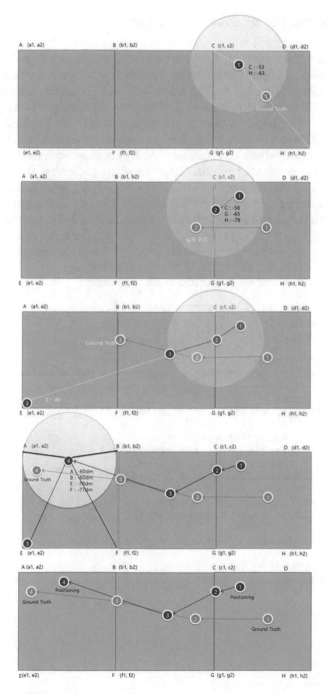

Fig. 7 Device positioning using plate technique

3.2 Pedestrian Dead-Reckoning Process

Pedestrian dead reckoning (PDR) is the process of calculating from a pedestrian's current position to a new position, using a previously determined position and predicting the next position based on known or estimated speeds over elapsed time and course. The PDR technique is naturally composed of three parts—step detection and estimation, step length estimation and heading estimation. Step counting, stride length estimation and heading estimation are enabled by inertial measurement sensors built into Android or iOS mobile devices. By continuously updating these three techniques, it is possible to estimate the route of the pedestrian. Inertial measurement sensors include an accelerometer for counting steps, a magnetometer using for figuring out heading, a gyroscope and a barometer. The 1st step is a low pass filtering process which removes noises of IMU sensor data, and the 2nd step is the calculation of a number of step using accelerometer and of heading defined with magnetometer.

Part of the program code for Inertial Measurement Units managing for the purpose of calibrating weaknesses of the accelerometer is as follows:

```
Private PzSensorManager(Context context)
{
mSensormanager = (SonsorManager)
context.getSystemService(Context.SENSOR.TYPE_ACCELEROMETE
R);

  accelerometer =
mSensormanager.getDefaultSensor(Sensor.TYPE_ACCELEROMETER
);
  megentometer =
mSensormanager.getDefaultSensor(Sensor.TYPE_MAGNETIC_FIEL
D);
  gravity = mSensormanager.
getDefaultSensor(Sensor.TYPE_GRAVITY);
  VECTOR = mSensormanager.
getDefaultSensor(Sensor.TYPE_ROTATION VECTOR);

  mRempData = new PzSensorData9);
  mCurrentData = new PzsensorData();

  mGyroOrientation = new GyroscopeOrientation(context);

  isCalibration = fales;
  calibrationValue = 0.0f;
}
```

The following is also a part of program code for distributed heading calculation to use PDR;

```
Static pubic void SquareBufferOne(float[] target, foat
values)
{
    Final float amplification = 200.0f;
    Float buffer = 10.0f;

    float Check = target[0] - values;
    if(math.abs(Check) >= 180)
    {
        if(Check < 0)
            target[0] += 360;
        Else
            Rarget[0] -=360;
    }


    target[0] += amplification;
    values += amplification;

    target[0] = (float)(Math.sqrt((target[0] * target[0] *
    buffer + values * values) / (1 + buffer)));

    target[0] -= amplification;
    values -= amplification;
}
```

Through the above 2 processes, the temporary coordinates are generated. As a last step, the temporary coordinate is calibrated again in order to get a more precise coordinate. The generated coordinate combines with Plate. Figure 6 shows PDR and Plate combined flow.

3.3 Plate and Pedestrian Dead-Reckoning Combination

The shape of the BLE plate is a circle with a radius of 5 m which is made through the process of BLE plate generation. This circle is the calibration standard used to determined whether PDR coordinates are accepted or corrected and adjusted. Therefore, only the PDR coordinates within a plate are used as correction objects. This means that it is a valid coordinate when the coordinates calculated in the PDR process are within the BLE plate circle. The PDR coordinates outside the circle should be corrected in the circle, that is, within the BLE positioning plate range. Figure 8 shows that PDR coordinates which are located outside of BLE plate circle are calibrated into the nearest circles (plates) (Fig. 9).

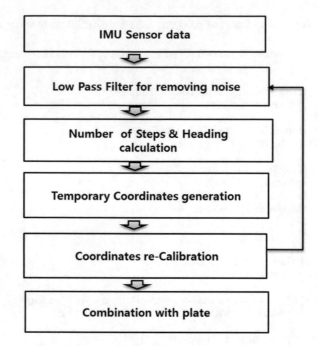

Fig. 8 PDR and plate combination flow

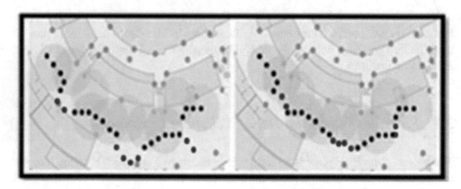

Fig. 9 Correction: coordinates outside of plate draw into plate circle

The following is part of the code for minimization of latency:

```
if(isSquareBuffer) {
    PzDataSmooth.SquareBufferforRound(mOrientationBuffer,
mCurrentData.getOrientation());
    PzDataSmooth.SquareBufferforRound(mGyroBuffer,mCurren
tData.getGyroscope());
}else
{
    mOrientationBuffer = mCurrentData.getOrientation();
    mGyroBuffer = mCurrentData.getGyroscope();
}

If(mSensorListener != null)
{
    PzSensorData.newData = new PzSensorData(mCurrentData);
    newData.setGyroscope(mGyroBuffer);
    newData.setCalGyroscopeCalVal(calibrationvalue);
    newData.setOrientation(mOrientationBuffer);
    m(sensorListener.onSensorChanged(new(Data);
}

    values -= amplification;
}
```

Figure 10 describes deviation of RSSI a hybrid estimation against actual moving path and a hybrid of Plate and PDR estimation which is actually tested at Gate 20 experimental site. The hybrid method is much closer to the actual path.

Figure 11 shows PDR logic combined with Plate technique. Top figure shows PRD estimate coordinate, RSSI hybrid coordinate and calibrated coordinate with plate technique against actual moving point. The 2nd figure indicates RSSI coordinate away from actual. PDR heading is similar to the actual direction, but far away from

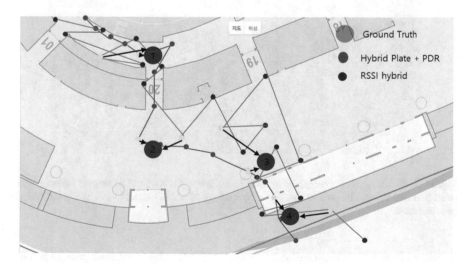

Fig. 10 Results of experiment with of RSSI hybrid only, plate and PDR hybrid against actual path

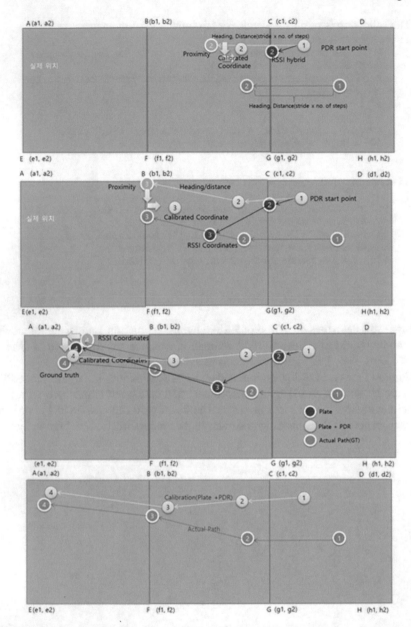

Fig. 11 Calibration process compared to RSSI plate. Hybrid Plate with PDR method calibrate next point to actual point. The bottom figure is a comparison of both Plate and PDR hybrid calibrated path and actual

the actual location. PDR coordinates, which calibrated by plate are closer to actual coordinate. The 3rd figure shows a position of the RSSI plate that is close to the

Beacon Map(1~120)   Beacon Map(501~545)   Beacon Map(546~625)

Fig. 12 Location of BLE beacons:125 beacons at 1st floor, 166 beacons at 2nd floor including corridor or crowded places

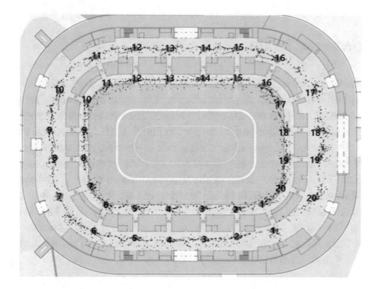

Fig. 13 40 access points were chosen as experiment points. The locations that were the busiest and had high levels of radio attenuation were selected to get results that reflected the real performance environment.

actual, but calibrated coordinate with hybrid plate and PDR is almost the same as the actual coordinate. The bottom is a comparison of calibrated path with hybrid method and actual path. Plate and PDR combined path is close enough to actual to be used for indoor navigation (Figs. 12 and 13).

3.4 Experiment Setup

Precision measurements were carried out using Android, Apple smartphone in ice sport arena which has various spaces, shapes and many communication barriers causing interference and propagation of radio. The double floor arena's space is

40,000 square meters with 8,000 seats for spectators. The electrical substation and the building control center are installed under the stadium, reducing the accuracy of the inertial measurement system.

125 BLEs were placed on walls, with cell at 1st floor and 120 beacons at the 2nd floor. 46 BLEs (501~546) were located in corridors or places with high crowding by spectators to improve the RSSI distance precision. Beacon place map data is located stored on the server in a database. Once the mobile device indoor positioning system is activated, map data is downloaded to the mobile device.

Real-time distance and location calculations were performed on the mobile device at 40 selected mostly busy locations as shown in the Fig. 13.

3.5 Experiment Results

Figure 14 shows the average of the data measured 10 times a day for 10 days, - that is, a total of 100 times at 40 test sites. The number represents the error distance in meter between the actual location and the estimated one. The average error distance of BLE plate is 6.1 m for the 1st floor and 6.3 m for the 2nd floor. Error numbers excessively larger than the average of each layer means that there are communication obstacles such as columns, walls and many movements of the crowd which are significant causes of reflection, interference and attenuation of radio waves. However, at point 10 at the 2nd floor, the error between the actual and the estimated position was reduced from 10.0 m to 1.4 m by applying the Plate and the PDR hybrid algorithm proposed in this paper. The results of the experiment using BLE Beacon Plate and PDR hybrid algorithm show that the average errors on the 1st and 2nd floor are reduced to 1.8 m and 2.6 m respectively.

1st Floor											
Point	BLE	Hybrid	Path	BLE	Hybrid	Point	BLE	Hybrid	Point	BLE	Hybrid
1	6.0	1.3	6	4.0	2.1	11	7.7	2.4	16	3.6	2.0
2	5.2	1.8	7	3.9	1.7	12	6.3	4.1	17	3.7	0.9
3	7.4	3.3	8	5.4	1.2	13	8.9	1.1	18	7.9	3.0
4	6.1	2.4	9	5.5	3.0	14	5.0	2.9	19	6.7	2.5
5	7.8	2.1	10	8.2	2.2	15	9.1	2.3	20	4.1	1.7

2nd Floor											
Point	BLE	Hybrid	Point	BLE	PLATE	Point	BLE	Hybrid	Point	BLE	Hybrid
1	7.4	2.0	6	9.4	3.3	11	5.3	2.5	16	5.7	2.2
2	5.9	2.7	7	5.2	2.1	12	8.8	3.8	17	9.1	1.5
3	4.0	2.9	8	6.8	1.8	13	6.5	2.9	18	7.1	1.2
4	8.0	1.3	9	4.1	1.1	14	6.1	2.3	19	4.9	2.6
5	10.0	1.4	10	4.1	0.7	15	3.0	3.0	20	5.0	2.1

Fig. 14 Comparison BLE and Hybrid Plate Algorithm with PDR at 40 experiment points

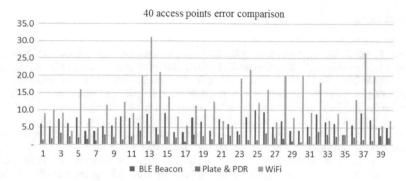

Fig. 15 Comparison of location estimation errors of BLE plate only, Plate and PDR hybrid and Wi-Fi

After applying the BLE Plate and PDR hybrid technique at each of the 40 experimental points, the graph in Fig. 15 shows an accuracy improvement of 71% on the first floor and 59% on the second floor. Both floor average error was improved from 6.2 m to 2.2 m at 40 places, which is 65% improvement.

Average error distance was determined to be 2.2 m in a very large two-story sports facility which accommodates over 8,000 spectators.

4 Conclusion

In this paper we propose a method to improve the accuracy of indoor positioning by combining BLE Plate and PDR. To increase the precision and verify the practicality, BLE Beacon was installed in an indoor space of a large building where many machines were installed and operated , rather than an artificial environment created for the experiment. The results of the study confirm that an accuracy of 2.2 m in a real indoor environment is suitable for a positioning system or indoor navigation for ultra-large spaces, such as airports and sport stadiums. This technique can be used in a seamless navigation through combining it with GPS and Wi-Fi for outdoor positioning and can also be applied to Augmented Reality services.

References

1. Chai, S., An, R., Du, Z.: And indoor algorithm using bluetooth low energy RSSI. AMSEE (2016)
2. Fang, L., Antsaklis, P.J., Montestruque, L.A., Brett McMickell, M., Lemmon, M., Sun, Y., Fang, H., Koutroulis, I., Haenggi, M., Xie, M., Xie, X.: Design of a wireless assisted pedestrian dead reckoning system—the navmote experience. IEEE Trans. Instrum. Measurement **54**(6) (2005)
3. Heinecke, T., Wolfe, M.: The Role of bluetooth low energy for indoor positioning application. In: Computer Science Department, Montana State University, Bozeman, MT USA
4. Huh, J., Lee, C., Kim, J.: A study of beacon delivery characteristics in BLE based fingerprinting indoor positioning system. Proc. KIISE **2015**(6), 1612–1614 (2015)
5. Kaemarungsi, K., Krushnamurthy, P: Modeling of indoor positioning system based on location fingerprinting, pp. 1012–1022. IEEE (2004). Telecommunication Program, School of Information Science, Unversity of Pittsburgh, Pennsylvania
6. Li, X., Wang, J., Lui, C: A Bluetooth/PDR integration algorithm for an indoor positioning system. Sensors (Basel), **15**(10) (2015)
7. Mautz, R.: Indoor positioning technologies institute of geodesy and photogrammetry, department of civil, environment and Geomatic engineering, ETH Zurih (2012)

A Study on the Common Collaboration Platform Activation of Cloud-Based Manufacturing Supply Management System (SCM)

Sanhae Kim, Hongjae Lee, Kyeong-Seok Han and Jong-Bae Kim

Abstract It is important to study and apply the application method developed to service from the SCM module for designing SaaS level to the cloud system, which supports SCM tasks such as procurement, purchase, logistics and standard information as to industrial sectors. These industrial sectors are divided into sewing clothes manufacturing, other metal processing product manufacturing and building construction business in terms of usage targets of cloud-based common collaboration platform. In addition, based on the importance and object of works such as common collaboration process, cloud-based common collaboration platform objective model is designed by selecting development model and service model suitable for this study. The cloud-based common collaboration platform will be used by many and unspecified companies, which is provided as the form of public cloud. Target model design means SaaS level finally. In the case of module development costs of collaboration works, it is required to be studied more to the future. By the way, in the case of the common collaboration module, the SCM tasks which can be applied to all industrial sectors have been calculated, which was based on the manufacturing industry. Since the SCM-related functions provided by the cloud-based common collaboration platform are not specific to the specified companies sectors, they are available to all companies. Furthermore, the SCM-related functions provided by the cloud-based common collaboration platform are various and selected by the modules freely because they are provided at the level of the cloud SaaS. The SCM-related functions provided by the cloud-based common collaboration platform are easy to

S. Kim · H. Lee · K.-S. Han
Department of IT Policy and Management, Soongsil University Graduate, School, 369,
Sangdo-ro, Dongjak-gu, Seoul, Korea 06978, South Korea
e-mail: james.kim@goodconsulting.co.kr

H. Lee
e-mail: hj1253@urpsys.com

K.-S. Han
e-mail: kshan@ssu.ac.kr

J.-B. Kim (✉)
Graduate School of Software, Soongsil University, Seoul, Korea, South Korea
e-mail: kjb123@ssu.ac.kr

© Springer International Publishing AG, part of Springer Nature 2019
R. Lee (ed.), *Big Data, Cloud Computing, Data Science & Engineering*, Studies
in Computational Intelligence 786, https://doi.org/10.1007/978-3-319-96803-2_3

select. It is necessary to make universal service available to the enterprise unlike the government that has been carrying out the support project to the specified company sectors on a one-to-one basis.

Keywords Cloud · SCM · Platform · Activation · Manufacturing

1 Introduction

The government has supported collaboration system consulting and implementation for various small and medium-sized companies by various type of business enhancing the degree of corporate interest and proving productivity improvement. However, there was a limit to the universal spread to the industry as a whole. Therefore, it is necessary to convert into a common cloud-based collaboration SW platform development and dissemination policy in order to spread the effect of the IT collaboration system to a large number of business groups. The internet is closely linked to our lives so that there are no places that have not affected the economy, culture and society of mankind. Accordingly, it is required to establish a software provision foundation that enables interaction between companies. It is necessary to emphasize mutual common function to support cooperation among companies, rather than supporting to use a specific SW to independent or closed companies groups. It is studied to have the possibility of selective usage by companies to utilize the functions provided by designing application systems for procurement, purchasing and logistics services as SCM (Supply Chain Management) selecting 3 industries to building construction industry, sewing clothes manufacturing and other metal processing product

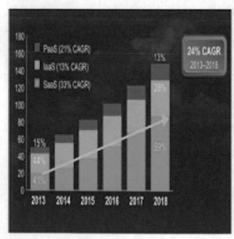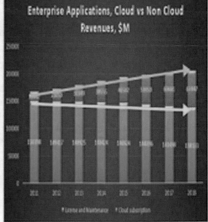

Fig. 1 Cloud solution introduction trend. The source of the left figure is Cisco global cloud index, 2013–2018. The source of the right figure is Apps Run the World, July 2014, which shows enterprise applications: cloud versus non-cloud revenues

Table 1 Major business status of overseas business software cloud SaaS

Major business	Offer service system, method, solution
salesforce.com	Cloud services are provided centered on CRM solutions. We can refer to Salesforce CRM as a typical SaaS application representatively. Currently, we are expanding our product range with marketing tools and manpower management (HR), etc. Signed a partnership with ABB, a leading power and automation technology company with over 145,000 employees in more than 100 countries, strengthening cloud service infrastructure (Sep., 2014)
ORACLE	We continue to expand our cloud application suite with in-house development and aggressive mergers and acquisitions and provide services integrating applications such as ERP, Human Resource Management (HRM) and customer experience (CX), etc.
Google	Google's representative SaaS services include enterprise business; Google Apps such as gmail, Google Calendar, Google Docs and Google Talk, etc., at a fraction of the cost. Currently, there are over 5 million corporate customers
SAP	SaaS has been strengthened by acquiring Ariba, which provides HRM (Human Resource Management) vendor SuccessFactors and a cloud-based provider network. In addition, we developed the business by design enterprise resource planning suite through internal development. We established SAP HANA Enterprise Cloud Center to LG CNS's Busan Data Center and strengthened collaboration to the field of cloud ERP
Microsoft	Accelerating SaaS promotion through its own dynamic ERP software suite, providing SaaS virtual one-stop services from office to CRM and ERP Representative SaaS services include Office 365, which is provided through MS Office and Cloud. Also, there are Dynamics CRM/ERP, etc. that are essential for corporate management

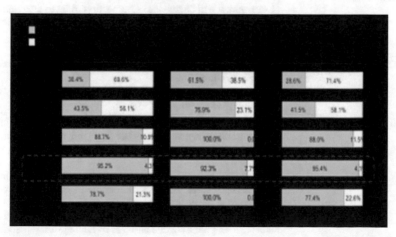

Fig. 2 Collaborative experience distribution in the field of SCM relative to the total value chain of Korean manufacturing [7, 5]

Table 2 Major business status of Korea's business software cloud SaaS

Major business	Offer service system, method, solution
Government agencies	To develop Korea's SW industry, development of specific SW & solution and free distribution SME Technology Information Promotion Agency develops software for business based on cloud computing through informatization business support for SMEs based on management innovation platform A total of 14 specialized solutions and 57 solutions for small businesses are provided free of charge on a cloud bases
DAOU Tech., INC.	We have focused on developing business applications optimized for the cloud environment. We have developed DAOU Office, a next-generation groupware solution that enhances the mobile work environment, collaboration and unified communication functions, TeamOffice, a team-based collaboration solution based on the cloud, Launched 'OfficeTalk' for enterprise SNS Collaborate with ERP specialist company, Younglimwon Soft Lab, to co-operate with DAOU Office and smart process
Hancom INC.	Designate SaaS services as core competencies of the company. The main products include ThinkFree, which provides office services through a web browser, and ThinkFree Mobile, which allows users to view and edit office documents while on the move using a smart phone or tablet, without having to install a separate office program on the user's PC Provide 'Netfice' which integrates web office, Hancom Office, e-Book authoring tool, etc. based on proprietary cloud platform technology to exploit SaaS market
Younglimwon Soft Lab	In the future, SaaS cloud service will be released in the form of SaaS, which will charge the customer according to the required service, instead of releasing the ERP package to the MS cloud 'Azure' It is estimated to account for 15% of total sales
DOUZONE ICT GROUP	We have cloud ERP products such as Smart A Cloud Edition and ICube Cloud Edition Attempt to increase sales through ERP 2 that incorporates cloud technology to existing ERP system
HANDYSOFT	Launched SaaS groupware with TILON INC., a virtualization specialist company, and introduced the Handypia SaaS platform a cloud service platform that provides a variety of applications online Handypia Platform: Unlike existing systems that provide cloud services on separate servers with multi-tenants, multiple users can access services through a single server We plan to conduct SaaS-type platform business and extend it to independent public service business by adding various services to future platforms Converting our collaboration solutions to SaaS-based for platform validation

(continued)

Table 2 (continued)

Major business	Offer service system, method, solution
TILON INC.	EL Cloud can be freely used only by connecting to the itnernet using cloud computing technology without installing various SWs such as various documents, graphic editing solutions and internet browsers directly on the handset
KongYoung DBM Co.	CRM Solution (MonArch) specializing company. Provided services such as customer management, campaign management, service management, sales management, reports, OLAP analysis & RFM analysis, etc. based on SaaS Compared to other vendors' solutions, it is advantageous for securing customer data reliability, providing flexibility for company-specific information management and providing convenience of work
KAON Internetworking	B2B co-operators for KT & SMEs/venture companies. Operation of Software as a Service method BizMeka Groupware service Interlocked product sales with UC and MS Exchange based enterprise applications such as KT Soip and KTF Mobile, etc.
Business on Communication Co., LTD.	As the first SaaS specializing compa; ny in Korea, it provides SaaS method to acquire solution effects, providing not solution method but service method to the fields of electronic tax invoice (smart bill, electronic contract and SCM)

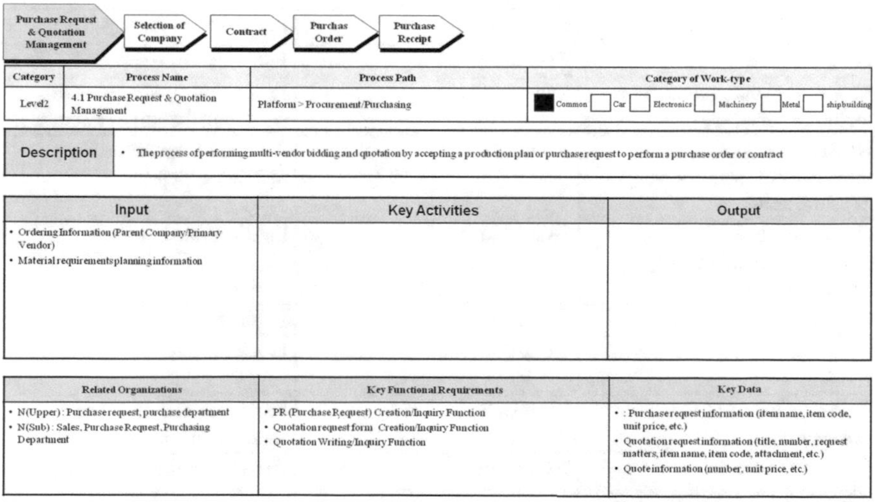

Fig. 3 Procurement/purchase process definition chart [8, 9, 5]

manufacturing industry. IaaS, PaaS and SaaS were designed to service the developed application system a s a cloud system. Especially, we have focused on SaaS. We have designed a roadmap and operation organization for implementing and operating the cloud system designed in this way and reviewed the roles of private companies and

Table 3 Table of the first classification result for selecting research object industry group

Main category	Selection/Exclusion	Remarks
A. Agriculture, forestry & fisheries	Exclusion	Individual Business
B. Mine	Exclusion	Collaborative structure is weak or large corporation type
C. Manufacturing	Selection	Leading public corporations such as power generation and transmission
D. Electricity, gas, steam & air conditioning supply business	Exclusion	
E. Water, sewage & waste disposal, raw material recycling	Exclusion	Large public corporation, small individual business
F. Construction	Selection	
G. Wholesale & retail	Exclusion	Simple distribution business with weak collaborative structure
H. Transportation & warehousing	Selection	
I. Accommodation & restaurant business	Exclusion	Individual business
J. Information & communication industry	Selection	
K. Finance & insurance	Exclusion	Large-scale industry group
L. Real estate	Exclusion	Individual Business
M. Professional, scientific & technical services	Exclusion	Professional industry with weak collaborative structure
N. Business facility management, business support & rental services	Exclusion	A public corporation type industry group
O. Public administration, defense & social security administration	Exclusion	Public affairs
P. Education services	Exclusion	Public affairs
Q. Health & social welfare services	Exclusion	Public affairs
R. Arts, sports & leisure services	Exclusion	Professional industry with weak collaborative structure
S. Associations & organizations, repair & other personal services	Exclusion	Group activities with weak collaborative structure
T. In-house employment activity, self-consumption production activity	Exclusion	Individual business
U. International & foreign institutions	Exclusion	N/A

Table 4 Result of the Second classification for selecting research object industry group

Industry classification	Company		Employees		Sales	
	Company no.	Ratio (%)	Employees no.	Ratio (%)	Sales	Ratio (%)
Sum total of whole industry	3,695,298		20,889,257		5,311,197	
Sum total of industry surveyed	944.047	0.03	7,512,050	35.95	2,498,661	47.05
(1) Sewing apparel manufacturing	19,989	2.12	130,674	1.74	33,801	1.35
(2) Other metal processing products manufacturing	50,486	5.35	333,794	4.44	64,821	2.59
(3) Building construction business	11,544	1.22	204,072	2.72	140,352	5.62

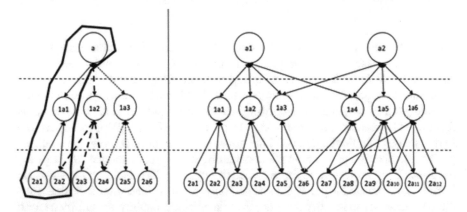

Fig. 4 Comparison between Existent IT collaboration system support method & common collaboration platform one [5] The left figure means the closed SCM. Oppositely, the right figure means the open SCM, cloud-based common collaboration model

governments. In addition, the costs of implementation and operation are separately estimated in terms of cloud operators and users companies. It emphasizes differentiation in terms of the existent government support policy in the fields of versatility, flexibility and ease of use. As qualitative expectations, it can standardize and systematize the sourcing process from purchasing request to contract to purchasing part. To procurement part, it is possible to realize order information of the mother company, synchronized production, delivery and receipt information management and enhance efficiency. Furthermore, in the case of the logistics service, it is predicted that the efficiency of the logistics management system is to be enhanced by establishing the logistics plan synchronized with the delivery request information of the parent company. On the other hand, the quantitative expectation effect requires the

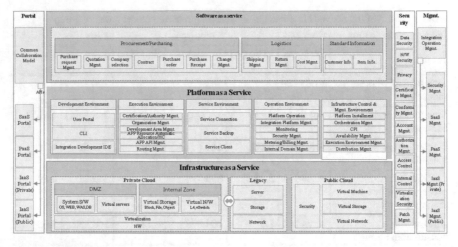

Fig. 5 Cloud-based common collaboration platform objective model [13, 14, 5]

initial development cost of KRW 32.5 billion and the annual operating costs of KRW 325 million in case of self development and operation, while the initial development costs are KRW 0 and the annual operating costs are KRW 18 million in case of the government support and cloud system leasing. In this study, we will present the reasonable grounds why the related government supports are required considering the fact that SMEs cannot afford to invest in IT by investing KRW 3.2 billion. Therefore, in this study, in particular, although the government has been carrying out a support project to a specific business sector considering the aspect of service, we hope that the universal cloud service can be provided through this study facing the reality that it is difficult to expand the policy since it was only provided for a specific business sector. It is necessary to identify the process of flexibility, reliability, versatility, security and interaction of the cloud-based common collaboration platform system, which leads to the intention and usage behavior of the system. Until then, it is necessary to proceed with the government initiative, and after that, the development of the private led type will fix the successful establishment of development and services (Figs. 1, 2, 3, 4, 5 and Tables 1, 2).

2 Theoretical Background

2.1 Cloud-Based Common Collaboration Platform

Cloud Computing is a service that pays the fee according to usage after borrowing IT resources such as HW, SW, etc. as much as necessary. Cloud Computing becomes important globally because it can reduce the cost of introducing systems and enable rapid IT system implementation as well as service delivery. For this reason, major developed countries are actively implementing policies by establishing mid and long-term plans so as to introduce them systematically recognizing the effects of Cloud Computing. In this study, the cloud-based common collaboration platform is to find the way to work smoothly between companies based on the cloud service. First, the functions that can be used to collaborate with each other should be provided as a computer program. Second, it should be provided on the cloud service so that these computer programs can be used selectively. In other words, the cloud-based common collaboration platform means a computer program based on the cloud service that enables companies to collaborate with each other using them in common, which can be said to mean SaaS among the tree cloud services [1, 2] (Tables 3 and 4).

2.2 Cloud-Based Common Collaboration Platform Implementation Case

SaaS will be the market leader in the cloud solutions market according to a recent report from U.S.A. HP. Enterprise applications such as enterprise resource planning (ERP) and customer relationship management (CRM) will become lead the Cloud SaaS (Software as a Service) market. It is expected that the cloud SaaS market will account for about 60% of the global cloud market by 2014 and grow at an average annual rate of 20% by 2017 [3–5].

Cloud SaaS is dominated by the big foreign companies such as SAP, Oracle, Microsoft and IBM. The leading companies are cloud companies, but traditional implementation software companies such as Oracle and SAP are also expanding their business. Among overseas cloud service providers related to this research, topic major market trends of business software SaaS are as follows.

In Korea, the central government had established the government integrated data center, but the reality is that it is still being reviewed by enterprises, local governments and public agencies. Of these, it should be noted that in the case of the public, the government integrated data center, which is responsible for the efficient management and stable operation of national information resources, is an example of the government integrated data center. It is in the process of implementing and ordering the project of implementing a cloud platform pilot service to reduce the introduction and operation costs of the information system and to provide rapid IT services to the service users.

IaaS, PaaS and SaaS can be used to implement a common collaboration platform based on the cloud. However, this study aims to explore how to collaborate with each other through specific application programs. Therefore, SaaS has been investigated. Looking at the major market trends of SaaS among the Korean cloud service providers, it can be seen that government agencies and software companies are offering or providing various solutions as follows.

2.3 Selection of Research Target Industries for Introduction of Common Collaboration Platform

The government has supported collaboration system consulting and implementation for small and medium-sized companies in various industries to enhance corporate interest and productivity improvement. However, the government has limited support for manufacturing industries such as automobiles, electronics, machinery, metals and shipbuilding. Accordingly, there was a limit to the universal spread of the industry as a whole. Therefore, in this study, it is necessary to investigate and select the Korean industry group fundamentally in order to spread the cloud-based common collaboration platform throughout the industry.

In order to select the industry group that is suitable for this study, we set the selection criteria of the industry group and set the industry group as a target to satisfy more than a certain level.

So as to select the industries to be studied, the population is based on the latest Korean Standard Industrial Classification of Korea National Statistical Office, which is adopted as the final selection by extracting the items appropriate for this study by industry group. However, the industrial groups such as automobile, electronics, machinery, metal and shipbuilding, etc. that has been supported to the past were excluded.

3 Selection of Target Industries

In this study, we divided the industry into two groups. In the first choice, based on hierarchical classification to the Korean Standard Industrial Classification, (1) industry groups that are in the form of private companies (excluding individual companies), (2) industry groups that can lead the market at the level of SMEs and (3) simple industry groups whose collaboration structure is remarkably weak are classified as exclusion, four industries including the first classification result were selected.

In the second category, the four industries selected in the first category are classified into 115 subcategories: (1) Industry group without government supports prior to this study, (2) Industry group with a strong collaboration structure, (3) Industry group with a distribution ratio of 2% or more to the total number of enterprises surveyed, (4) Industry group with a distribution ratio of 1.5% or more to the total surveyed population and (5) industry group with a distribution ratio of 1.2% or more of the total sales.

As a result of analyzing 115 industries according to the second classification standard, sewing clothes manufacturing, other metal processing products manufacturing and building construction industry were analyzed and selected as follows [6, 5].

3.1 Analysis of Common Collaboration Task Survey Applied to Many Companies in Industry

The re-design was performed by extracting the tasks that can be used among actual companies and expected to be effective. According to the results of the previous research, it is understood that the cooperation rate is declining from the viewpoint of supply chain network to the second to third SMEs, and that SMEs need collaboration in order/revenue collaboration, information sharing and joint technology development.

In the previous research, the target group of companies was the parent company and the first partner company, which deals with the parent company's way of handling the business relationship with the first partner. As a result, it can be said that it is a kind of business process that is entrusted to a specific company. In brief, this study does not fit the objective of this study because it is not universal in the horizontal direction. In this study, we conducted a study on the common platform for the second to nth companies of the lower level based on the first partner.

Among the collaboration tasks that can be done among the enterprises, the analysis of the other items of the systems used by the parent company and the partner companies shows that the proportion of collaboration is high to procurement, purchasing and logistics. Even though it is estimated that there is a somewhat different situation between the parent company (large corporations) and the first partner and the second to nth partners, in the case of manufacturing industry, it is judged that collaboration experience and necessity in procurement, purchasing and logistics services will co-remain.

SCM is a system in which all companies participating in the distribution supply chain, such as manufacturing, logistics and distribution companies utilizing information technology to optimize inventories based on cooperation and drastically reducing lead time as well as ultimately reducing production costs and increasing sales.

In this study, we designed the business that can achieve the highest effect among the businesses that can collaborate with each other to the extent that cloud service is possible. The method of deriving the common collaboration service refers to the basic functions of SCM (purchasing/procurement/logistics) and previous research results through visits and surveys to the first to nth companies so as to set up the most common and easy-to-use functions. The process of procurement/purchasing business can be divided into purchase requisition and quotation management, company selection, contract exchange, purchase order and purchase receipt. Each activity is organized into activities, which describes in detail the items necessary for future functions, data and screen configuration subdividing each process such as a purchase request.

A cloud-based common collaboration platform for supporting SME/s purchasing/procurement/logistics businesses shall be developed considering two prerequisites. In order to develop on the basis of the first condition, Cloud SaaS, a basic understanding of SaaS application performance is required. [3, 10, 9] SaaS allows users to utilize computing resources of remote sites without limitation of time and place based on the network, paying service costs only as they use them, which enables to respond immediately to users using virtualization technology. Because SaaS-based software pays a fee for each tenant, it is important to develop multi-tenant based services. The process of developing reference architectures and components that satisfy all requirements of multi-tenants is important and should be subdivided into multi-tenant requirement analysis, reference architecture design and component development. [5] This ensures stable profit by satisfying multi-tenant of various customers with only one instance (Single Instance).

Therefore, in order to develop based on cloud SaaS, the following should be contained. SaaS cloud service means a model that is deployed and run on a cloud platform so that users can use software as an online service. Customers who use SaaS cloud services are less burdened with initial investment or systems management. However, they should pay a fixed amount according to service period or usage. These SaaS cloud services are transforming existing IT environment from supplier-oriented to buyer-oriented secured the marketability by small-scale specialized solutions vendors. It is important to enable to deploy various types of SaaS cloud services according to the tenant's unique requirements because the development of SaaS cloud service requires a system that can support the development of services that can be easily customized according to the requirements of various tenants.

In order to determine the price as to the quantity and quality of the services used in the cloud service provision, a system for users to pay is required. We define an effective model that can provide a variety of services by paying as much as they use the IT service system based on usage. Of course, the charge for the user is subject to the charge policy of the provider. However, the design of the tenant is important in order to rationalize the charge policy. Tenants shall subdivide possible functions. In other words, the user should be provided with a function to utilize necessary functions selectively.

The security zone means independent security for the user's data area and access by shared or unauthorized users must be controlled.

Cloud providers provide applications built on a specific application platform, where the application platform uses a data platform hosted on the system infrastructure. Since the computing resources underlying the business application are hierarchical, the application's multi-tenancy can be implemented in any layer or hierarchy of the lower layer.

4 Conclusion

The purpose of this study is to support internet-based transactions for many unspecified SMEs, rather than targeting the specific business groups supported by the government. Therefore, each information system function derived from this study considers general purpose to be utilized by all companies that want to use this information system.

The cloud-based common collaboration platform is an open platform for developing a common system module (SW) for collaboration work that is applied to a large number of companies among large-to-medium enterprise collaboration processes and for recycling when introducing a collaboration system. It is possible to implement ICT collaborative environment with minimal costs and to provide ease of deployment, scalability and compatibility with the development of a small and medium-sized business partner who is dealing with a large number of parent companies. Prevention of system's redundant investment and relieving burden of system operation to small and medium sized enterprises lacking IT manpower are possible.

However, cloud-based common collaboration platform to support SCM (purchase/procurement/logistics) works of SMEs is technically to be possible to utilize computing resources of remote place without limitation of time and place based on network by users, enabling immediately to respond to user's needs using virtualization technology, that pays only for the service users use.

In the case that the SCM business is provided to the enterprise by using the common collaboration platform based on the cloud, the purpose of the information system function is to support the internet-based transaction for many unspecified SMEs. Consideration should be given to maintain generality so that companies can utilize them. The SCM model, which is the subject of this study, uses the same business process rules and the information system functions that have been developed for all contracting companies and all cooperating companies such as subcontracting and delivering. In other words, all companies should be able to conduct multiple transactions in accordance with common transaction rules such as business procedures and document rules and all transaction rules can only reflect parts of the ordering company and partner companies that satisfy the transaction processing conditions [11, 12].

In order to do this, it is necessary to support each company's work as much as possible, but it has to be accompanied by standardization and designed so that general-purpose can be maintained. However, if the emphasis is on general-purpose only, each function gradually becomes generalized so that the utility of each company may be deteriorated. Therefore, various functions have to be developed as much as possible in order to minimize inconvenience or constraints of each company's SCM activities. Therefore, there is a need for further research to be carried out.

In addition, the SCM common collaboration function provided by the cloud-based common collaboration platform is basically provided by SaaS. To accomplish this, each function is configured as multi-tenancy and provided to each user charging according to usage. This multi-tenancy structure has some problems including: (1) possibility of expanding program bugs and range of disabilities and (2) security problems etc. so that research on ways to cope with future technical issues needs to be conducted in depth.

References

1. Bartsch, V., Ebers, M., Maurer, I.: Learning in project-based organizations: The role of project teams' social capital for overcoming barriers to learning. Int. J. Project Manage. **31**(2), 239–251

(2013)
2. Makhija, M.: Comparing the resource-based and market-based views of the firm: empirical evidence from Czech privatization. Strateg. Manag. J. **24**(5), 433–451 (2003)
3. Llorens, S., Cifre, E., Martínez, I.M., Schaufeli, W.B.: Perceived collective efficacy, subjective well-being and task performance among electronic work groups an experimental study. Small Group Res. **34**(1), 43–73 (2003)
4. Srivastava, A., Bartol, K.M., Locke, E.A.: Empowering leadership in management teams: Effects on knowledge sharing, efficacy, and performance. Acad. Manag. J. **49**(6), 1239–1251 (2006)
5. National IT Industry Promotion Agency. A Study on the Common Collaboration Platform Activation of Cloud-based Manufacturing Supply Management System (SCM) (2018)
6. Bardhan, I., Krishnan, V.V., Lin, S.: Team dispersion, information technology, and project performance. Prod. Oper. Manag. **22**(6), 1478–1493 (2013)
7. Zaheer, A., Bell, G.G.: Benefiting from network position: firm capabilities, structural holes, and performance. Strateg. Manag. J. **26**(9), 809–825 (2005)
8. Brettel, M., Mauer, R., Engelen, A., Küpper, D.: Corporate effectuation: entrepreneurial action and its impact on R&D project performance. J. Bus. Ventur. **27**(2), 167–184 (2012)
9. Yu, Y., Hao, J.X., Dong, X.Y., Khalifa, M.: A multilevel model for effects of social capital and knowledge sharing in knowledge-intensive work teams. Int. J. Inf. Manage. **33**(5), 780–790 (2013)
10. Tasa, K., Taggar, S., Seijts, G.H.: The development of collective efficacy in teams: a multilevel and longitudinal perspective. J. Appl. Psychol. **92**(1), 17–27 (2007)
11. Tsai, W., Ghoshal, S.: Social capital and value creation: The role of intrafirm networks. Acad. Manag. J. **41**(4), 376–464 (1998)
12. Gu, V.C., Hoffman, J.J., Cao, Q., Schniederjans, M.J.: The effects of organizational culture and environmental pressures on IT project performance: A moderation perspective. Int. J. Project Manage. **32**(7), 1170–1181 (2014)
13. Gulati, R., Sytch, M.: Does familiarity breed trust? Revisiting the antecedents of trust. Manag. Decis. Econ. **29**(2–3), 165–190 (2008). Brettel, M., Mauer, R., Engelen, A., Küpper, D.: Corporate effectuation: entrepreneurial action and its impact on R&D project performance. J. Bus. Vent. **27**(2), 167–184 (2012)
14. Hackbarth, G.: The impact of organizational memory on IT systems. AMCIS 1998 Proc. **197** (1998)
15. Adler, P.S., Kwon, S.W.: Social capital: prospects for a new concept. Acad. Manag. Rev. **27**(1), 17–40 (2002)
16. Bandura, A.: Editorial. Am. J. Health Promot. **12**(1), 8–10 (1997)
17. Chang, H.H., Chuang, S.S.: Social capital and individual motivations on knowledge sharing: Participant involvement as a moderator. Inf. Manag. **48**(1), 9–18 (2011)
18. Chow, W.S., Chan, L.S.: Social network, social trust and shared goals in organizational knowledge sharing. Inf. Manag. **45**(7), 458–465 (2008)
19. Chen, M.H., Chang, Y.C., Hung, S.C.: Social capital and creativity in R&D project teams. R&d Manag. **38**(1), 21–34 (2008)
20. Coleman, J.S.: Social capital in the creation of human capital. Am. J. Sociol. **94**, 95–120 (1988)
21. DeRue, D.S., Ashford, S.J.: Who will lead and who will follow? A social process of leadership identity construction in organizations. Acad. Manag. Rev. **35**(4), 627–647 (2010)
22. Lee, J., Park, J.G., Lee, S.: Raising team social capital with knowledge and communication in information systems development projects. Int. J. Project Manage. **33**(4), 797–807 (2015)
23. Liao, S.H., Fei, W.C., Chen, C.C.: Knowledge sharing, absorptive capacity, and innovation capability: an empirical study of Taiwan's knowledge-intensive industries. J. Inf. Sci. **33**(3), 340–359 (2007)
24. Liebowitz, J.: Knowledge management and its link to artificial intelligence. Expert Syst. Appl. **20**(1), 1–6 (2001)
25. Meng, X.: The effect of relationship management on project performance in construction. Int. J. Project Manage. **30**(2), 188–198 (2012)

26. Molina-Morales, F.X., Martínez-Fernández, M.T.: Social networks: effects of social capital on firm innovation. J. Small Bus. Manage. **48**(2), 258–279 (2010)
27. Nahapiet, J., Ghoshal, S.: Social capital, intellectual capital, and the organizational advantage. Acad. Manag. Rev. **23**(2), 242–266 (1998)
28. Di Vincenzo, F., Mascia, D.: Social capital in project-based organizations: Its role, structure, and impact on project performance. Int. J. Project Manage. **30**(1), 5–14 (2012)
29. Stajkovic, A.D., Lee, D., Nyberg, A.J.: Collective efficacy, group potency, and group performance: meta-analyses of their relationships, and test of a mediation model. J. Appl. Psychol. **94**(3), 814–828 (2009)
30. Tsai, Y.H., Ma, H.C., Lin, C.P., Chiu, C.K., Chen, S.C.: Group social capital in virtual teaming contexts: a moderating role of positive affective tone in knowledge sharing. Technol. Forecast. Soc. Chang. **86**, 13–20 (2014)
31. van Emmerik, H., Jawahar, I.M., Schreurs, B., De Cuyper, N.: Social capital, team efficacy and team potency: The mediating role of team learning behaviors. Career Dev. Int. **16**(1), 82–99 (2011)
32. Wang, S., Noe, R.A.: Knowledge sharing: A review and directions for future research. Hum. Resour. Manag. Rev. **20**(2), 115–131 (2010)
33. Wah, C.Y., Menkhoff, T., Loh, B., Evers, H.D.: Social capital and knowledge sharing in knowledge-based organizations: an empirical study. Int. J. Knowl. Manag. **3**(1), 29–48 (2009)
34. Hoegl, M., Weinkauf, K., Gemuenden, H.G.: Interteam coordination, project commitment, and teamwork in multiteam R&D projects: a longitudinal study. Organ. Sci. **15**(1), 38–55 (2004)
35. Liu, Y., Keller, R.T., Shih, H.A.: The impact of team-member exchange, differentiation, team commitment, and knowledge sharing on R&D project team performance. R&D Manag. **41**(3), 274–287 (2011)
36. Gu, Q., Wang, G.G., Wang, L.: Social capital and innovation in R&D teams: the mediating roles of psychological safety and learning from mistakes. R&D Manag. **43**(2), 89–102 (2013)
37. Westerman, G., Bonnet, D., McAfee, A.: Leading Digital: Turning Technology into Business Transformation. Harvard Business Press (2014)

Service Management and Model Driven Management

Haeng-Kon Kim

Abstract The traditional model management and the service system, which is the key element of DSS, only support the single format model in the special application field. Based on the decision requirements assisted by several models, this article designs the general model and the service management system that provides general administration and service for different models of format in the network. The system proposes the solution to manage different format models by providing a framework generator model that aims to analyze the top-down framework of the decision problem and the application of the model that aims to integrate the bottom-up model function.

Keywords Decision support system · Model-aided decision · Model base
Model management

1 Introduction

Model management, one of the key elements of the decision support system (DSS), has been widely studied and has attracted a lot of attention. Researchers have designed several representations of models based on advanced software and hardware technologies, and implemented numerous systems that are used for model management, composition, and model services. Although the series of significant productions has provided model-assisted decision-making solutions at this time, there is virtually no administration and service for the different format models in the network. As the information industry grows, model-assisted decision-making receives more and more attention. The larger numbers of resources of the basic models and business models are the basis for model-assisted decision-making, while posing an unprecedented challenge, for example, since models are developed and reconstructed in different technological architectures, which causes distributed locations and heterogeneous techniques. However, few solutions are effective to manage model resources with

H.-K. Kim (✉)
School of Information Technology, Daegu Catholic University, Gyeongsan, South Korea
e-mail: hangkon@cu.ac.kr

© Springer International Publishing AG, part of Springer Nature 2019   49
R. Lee (ed.), *Big Data, Cloud Computing, Data Science & Engineering*, Studies
in Computational Intelligence 786, https://doi.org/10.1007/978-3-319-96803-2_4

different formats in the network. The model's resources as fixed terms in the control process, which are developed for a special application and are used in some way, are not shared and used sufficiently. Therefore, it is necessary that the model and service administration system manages models with different formats in the network and offers several different solutions for sharing models. As part of the model-assisted decision requirements, this document designs and implements the general model management and service system that manages multiple models compiled in different formats, such as DLL, COM, EXE and web services, and provides two support mechanisms. One is the top-down decision problem and the other is the bottom-up integration of functions. The system provides the solution for sharing and applying heterogeneous decision models [1–4].

2   Formal Description of Model

Programmers must write the code processing the details of the calculation, synchronization, heterogeneity and communication systems. Programmers must also use low-level language constructs to implement these details. In general, it makes distributed programming more complicated and prone to errors [5]. The information in the description of the model that forms the basis of the management of the model and the shared services must be descriptive, practical and detailed. There are three levels to describe the model information: syntax, semantics and pragmatics.

2.1   Model Pragmatics Information

The pragmatic information model about the classification of the model, the function of the model and the version contains the following information:

- model name;
- model information about the application domain and classification, by which models are describe from several different aspects;
- model function;
- model version.

2.2   Model Semantic Information

The semantic model information that is related to the model principle, the development statement, the control results and the interface parameter information contains the following information:

- model principle;
- model instruction of the development;
- developers information;
- version modified information;
- authorization information;
- verification information;
- information of methods provided by model and whose function;
- information of in-out parameters;
- Information about files associated the model such as data files, help files, source files, reference files, temp files, etc.

2.3   Model Syntax Information

The syntax information about the model on the formats, orientations and grammatical structure of the input and output parameters of the model contains the following information

- information about environment under which models are developed and running;
- model formats;
- location information about the model resources such as IP and port of the model service provider, file path, web services path and so on;
- information about how to invoke the methods of the model such as class name, method name, the count of parameters and so on;
- in-out parameters information such as ID, input or output tag, name, type, length, precision, default value, max value, min value and so on [6–8].

3   Systematic Design

In terms of basic principles and design philosophy, centralized administration, unified service and flexible configuration, the general administration and service system is divided into four tools: model description tool, model information management tool, application tool of three nodes in the network: model information management node, model service node and mode application node. Figure 1 describes the system that is configured flexibly according to the real environment and the situation: (1) the database servers are configured in the separate node or are connected to other nodes; (2) the minimum system that nodes that have been integrated into a computer can find; (3) There may be multiple model application nodes and model service nodes in a system.

(1) Model description tool. Semi-automatically converts template resources such as DLL, COM, EXEs, and web services into standard template information description files in the template service node, and automatically generates a

Fig. 1 System structure diagram

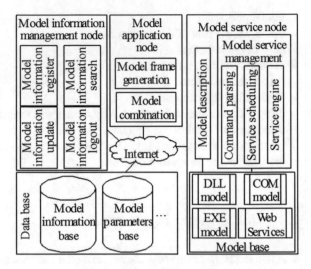

template resource based on a standard managed code that forms the basis of the unified model service.

(2) Model information management tool. It is located in the information node of the model, whose main functions are the following: (1) receive the annotated files in the model description information of the model service nodes and save the description information of the standard model in the base of model information; (2) disconnect the information from the model description from the information base of the model in a passive or initiatory manner; (3) search and select model description information from model information based on user needs; (4) update the model description information by sending requirements from the model service nodes; (5) to classify and apply models, build and modify the classification system dynamically.

(3) Template application tool. It provides two mechanisms to support model-assisted decision making in the model application nodes: (1) when the model functions converge, the tool allows assemble the models available in combined models to solve the decision problem. (2) When analyzing decision problems, the tool allows to disassemble the decision problems gradually into several interactive meta-problems associated with the available models, then automatically generates a framework of models combined with decision problems. The first mechanism can generate more flexible and abundant model sources based on real models available; the second mechanism is in accordance with the human habit of thought. These mechanisms complement each other and bring out the best in each one. The main functions of this tool are: (1) it provides the visible editor for the control flows and the information flows of the combined models; (2) Automatic generate a sequence of commands to control the combination patterns; (3) provides the mechanism for interpreting and sending service requests from metamodels to associated model service nodes. (4) template service tool.

It is located in the model service node, whose main functions are the following: (1) listens and receives model service requests from the nodes of the model application; (2) analyzes the service requests of the models, and as parameterization of the input and output parameters of the current models, constitutes the standard operating orders of the models; (3) manages the execution of the model by the service engine and records the results and records of the models in execution; (4) Depending on the flow direction of the output parameters, it sends the results of the execution to the application nodes of the model [9, 10].

4 Key Technology

4.1 Generation of Managed Code

There are several operators of the models by which the decision is supported. The formats supported by the following system:

(1) the standard dynamic link library (DLL);
(2) COM (Component Object Model) whose format is DLL or EXE;
(3) the standard executable program (EXE);
(4) Web services. On the one hand, it is the basis of centralized management to describe the model of the standard, which is also the first step in the management and service of the model. The system must adapt to automatic scans and extract information from the template, particularly the syntax information of the template, to guarantee the template description information in a uniform and practical standard for saving templates for users. On the other hand, since the modes of invocation of the model of the four operators are different, it is necessary to generate an intermediate code that is identified in the standard service interface to implement the principles, centralized management. The reflection mechanism that is the key feature of the .NET Framework is used to obtain information about the typical .NET member, such as information about methods, attributes, properties, events and constructors, etc., and to create and invoke the instances by the metadata The system implements the function to generate automated managed code and retrieves the description information of the model using the reflection mechanism and some Visual Studio .Net tools (such as TlbImp.exe, dumpbin.exe and wsdl.exe), and then executes the principle, centralized administration and unified service. It is necessary that each of the operators of the models uses the corresponding tool to generate the managed code and extract the information from the model. Figure 2 shows the process:

(1) COM Model: extracts and analyzes COM file information using Tlbimp.exe, and then transforms them into an assembly in the .NET Framework.

Fig. 2 Manage code
generation and model
extraction

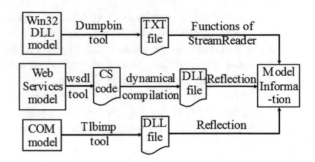

(2) Dynamic link library model: extracts and analyzes the information in the
DLL file using Dumpbin.exe and then transforms it into an assembly in the
.NET Framework.

(3) Web Services Model: extracts and analyzes information from the WSDL,
XSD or .discomap file using Wsdl.exe and then transforms it into an assembly in the .NET Framework.

The process of generating the corresponding managed code using the reflection
technologies is as follows:

(1) Define and load the assembly, then create an instance of Assembly type.
(2) Get the available classes and the global assembly methods by module type.
(3) Get the constructor information, such as name and parameters, and so on, and
then create an instance using the GetConstructors method or the Type GetCon-
structor method.
(4) It obtains the information of methods such as name, parameters and type of
return value, etc., and then obtains the methods by means of the GetMethods
method or the GetMethod method of the MethodInfo type.
(5) Obtain the event information, such as the name, procedure and corresponding
type, and so on by the EventInfo type and add or remove the event handler.
(6) Obtain information about the property, such as name and type, etc., then set or
get a value with the PropertyInfo type.
(7) Obtain parameter information such as name, type and location, etc., by type
ParameterInfo.

4.2 Model Application Frame

The initial phases of software product development require a careful analysis of the
business to which the application is addressed [11]. The problem of the decision is
often the complex problem formed by a series of sub-problems and that can only
be solved by multi-models. Two template application frameworks are provided in

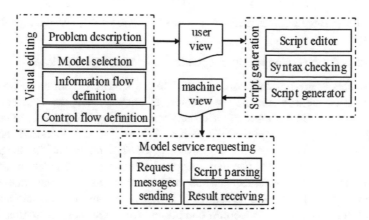

Fig. 3 Model application frame

the system. (1) Intended for the convergent functions of the model, the tool allows assembling the models available from the bottom up in combined models to solve the decision problem. (2) Intended for the analysis of decision problems, the tool allows disassembling the decision problems from top to bottom, and then automatically generates frame combination models. These functions are implemented according to the principle that the operation interface and the flow remain online. Figure 3 describes the tool.

(1) Visual editing module. It is the visual editor for assembling models and dis-assembling decision problems, through which the flow of information and the flow of control are edited. The properties of the decision problems and the com-bination models are described in the sub-module describing the problem. The relationships between sequence, selection and circulation of sub-problems or sub-models are visually edited to generate the control flow that is consistent with the solution of the decision problems by the control flow definition sub-module. Since the model selection submodule shows information from the standard and descriptive model, which assumes that the selection is dominated by the user and the message is assisted by the system, users can select the appropriate templates. The relationships of the input and output parameters are defined by the infor-mation flow definition module.

(2) Scripts generation module. The essential part of the generated scripts is to use the information and the relationship of the meta-problems to describe the user's view. In the process, it is necessary to verify if the parameters and the configura-tion of the models and the combined models are logical and integrated, and then the EBNF-based format generates .XML script files that are logical and clear and can be validated. The template script is provided for sophisticated users, which is indicated in Refs. [10, 11].

(3) Service request module. It sends service request messages that are generated according to the flow of control and information in the machine view script

and the parameter information to the model service node, and then receives the results.

The rules for designing and implementing the system are as follows.

(1) The rules of the standard control flow. The three basic structures between sub-models or subproblems are the structure of the sequence, the structure of the selection and the structure of the circulation according to the theory of structuring studied by G. Jacopini and C. Bohm. (1) Structure of sequence. This means that sub-models and subproblems are executed in sequence and that the parallel structure can be transformed into a sequence structure. (2) Circulation structure. This means that sub-models and subproblems repeatedly execute under certain conditions. (3) Selection structure. This means that sub-models and subproblems are executed under certain conditions. Although these structures can be nested theoretically, the nested level must be simplified.

(2) The rules of the preprocessing function of the modeled data. The type, length and precision of the input and output parameters are different from each other in the process of transmitting parameters. Therefore, it is necessary to implement data preprocessing such as data transformation and data cleaning, etc. in the process of defining the flow of information. The loading of the visual editing module is lightened and the problem solving procedure becomes legible and transparent when modeling the data preprocessing function.

(3) The rules of the parameters inside-out. The type of parameters expressed by the connection string is restricted as a type of string, and the read and write operations are performed by the templates, since the template transfer parameters through the database. The type of parameters expressed by the data file path is restricted as a string type, and the file operations are executed by the models, when the models transfer the parameters through the data file. The parameters are necessary for the transformation into a data or database file when the models transfer parameters through the multidimensional matrix.

5 Conclusion

Model-assisted decision-making is the key to MAS. Based on the current available technology system, the general model management and service system is implemented in this article by the author, who provides functions for different format templates to manage and maintain the network and analyze, automatically extract and manage the information of the template.

The two application frameworks of the system model are provided, namely the model assembly functions and the generation of model frames. As the increase of the model resource, the decision problem becomes more complex. In future work we plan to study how to select the automatic model and the exchange model in a hybrid environment.

Acknowledgements This Research was supported by the MSIP (Ministry of Science, ICT and Future Planning), Korea, under the ITRC (Information Technology Research Center) support program (IITP-2018-2013-0-0087) supervised by the IITP (Institute for Information and Communications Technology Promotion).

References

1. Yunus, M.: Creating a World without Poverty: Social Business and the Future of Capitalism. Public Affairs, p. 320 (2009). ISBN 978-1-58648-667-9
2. Social Business Architecture: IBM Corporation 2014. Retrieved from, 10 March 2015
3. Cheesman, J., Daniels, J.: UML Components: A Simple Process for Specifying Component-Based Software. The Addison-Wesley Object Technology Series 2000. ISBN: 0-201-70851-5
4. The Social Business: Advent of a new age, White Paper, EPW14008-USEN-00, Retrieved from https://www.ibm.com/smarterplanet/global/files/us__en_us__socialbusiness__epw14008usen.pdf, 10 February 2015
5. Mahmood, S., Ahmed, M., Alshayeb, M.: Analysis and evaluation of software artifact reuse environments. Int. J. Softw. Innov. (IJSI) **2**(2), 54–65 (2014)
6. Agner, L.T.W., Soares, I.W., Stadzisz, P.C., Simao, J.M.: Model refinement in the model driven development context. J. Comput. Sci. **8**(8), 1205–1211 (2012). ISSN 1549-3636
7. Cheesman, J., Daniels, J.: UML Components, A simple Process for Specifying Component-Based Software. The Addison-Wesley Object Technology Series (2001)
8. Cepa, V.: Product-Line Development for Mobile Device Applications with Attribute Supported Containers. Ph.D. Dissertation, Software Technology Group, Technical University of Darmstadt (2005)
9. Kim, H.-K.: Mobile agent development with CBD on ABCD architectures. In: International MultiConference of Engineers and Computer Scientists (2013)
10. Kim, H.-K.: Design of web-based social business solution architecture based on CBD. Int. J. Softw. Eng. Its Appl. **9**(2), 271–278 (2015)
11. Berrocal, J., Garcia-Alonso, J., Murillo, J.M.: Modeling business and requirements relationships to facilitate the identification of architecturally significant requirements. Int. J. Softw. Innov. (IJSI) **2**(1), 9–24 (2014)

Measuring the Effectiveness of E-Wallet in Malaysia

Faisal Nizam, Ha Jin Hwang and Naser Valaei

Abstract In view of the promising growth of E-wallet in Malaysia, this study aims to discover the important factors influencing consumers purchase decision using E-wallet. Previous studies reveal that factors such as convenience, security, and cost saving influence consumers' purchase decision using E-wallet. The survey questionnaire was developed and distributed to 230 respondents, out of which 222 valid responses were used for further statistical analysis. The result of this study indicated that convenience, security, and cost saving were proved to make significant influences on consumers purchase decision using E-wallet. The small sample size could limit the opportunity of generalization of the findings in this study, which future studies should seek to overcome. For practical implications, the use of E-wallet by the majority of respondents confirms that there is a great potential for future expansion of such payment devices in Malaysia.

Keywords E-wallet · E-payment · e-Commerce

1 Introduction

The E-Wallet provides users with the technological convenience of purchasing products or transferring funds to and from anywhere around the world. The E-wallet is a transaction system where an internet service or program allows consumers to handle information in a central place related to purchases, loyalty, membership, and banking information which is somewhat like physical wallet [17]. According to Nielsen-PayPal Analysis (2013) e-Commerce of Malaysia was estimated to be

F. Nizam · H. J. Hwang (✉) · N. Valaei
Sunway University, Subang Jaya, Malaysia
e-mail: hjhwang@sunway.edu.my

F. Nizam
e-mail: Faisalnizam1004@gmail.com

N. Valaei
e-mail: naserv@sunway.edu.my

© Springer International Publishing AG, part of Springer Nature 2019
R. Lee (ed.), *Big Data, Cloud Computing, Data Science & Engineering*, Studies
in Computational Intelligence 786, https://doi.org/10.1007/978-3-319-96803-2_5

around RM 24 billion by the year 2017 and 60% of e-Commerce would be done by mobile commerce.

As e-Commerce is rapidly developing in the world especially in the recent years, many aspects of life is affected by e-Commerce, especially the advancement in how people manage themselves financially and non-financially in various transactions. And ultimately, processes and systems dealing with this emerging trends would have to change for the betterment of efficiency and effectiveness. According to Bank of America (2014), E-Payables tries to transform the paper checks into electronic payments by reducing the usage of many manual aspects of account payable. This has led to the importance of the cost-saving or reduction in cost for organizations and consumers. Another benefit of e-payment is convenience. Chauhan [3] states that customer's data can be stored in E-wallet system for easy retrieval for online purchases. Besides, Monika (2006) mentions that this system also provides ease to customers in the online shopping process where the system already has all necessary transaction details. This then contributes to the justification of Convenience factor.

Moreover, Niranjanamurthy and Dharmendra [10], mention that theft and fraud, unauthorized access, and denial of service are the main security threats faced by the consumers while using e-payment system. This will cause consumers to avoid from using this system. Niranjanamurthy and Dharmendra [10] also suggest the solutions to prevent or minimize such threats need to be developed. This leads to the importance of security factor to be included in this research. This research is focused on the consumer purchasing behavior by measuring the effectiveness of E-Wallet system in Malaysia. The consumer market in Malaysia is still in the transition process towards an emerging market and e-Commerce system is still under development. It was noted by United Nations Statistics Division that the majority of consumers in Malaysia, with 29.6% under the age of 15, 65.4% between 15–64, and 5% ages above 65 are still using physical cash. Issues of why the Malaysian people have not adopted E-Wallet favorably can open up the opportunity to develop more efficient and user friendly E-Wallet system.

There has been little researches done to determine what and how Malaysians think of the E-wallet system and whether they are keen to use it daily. This study is designed to identify important factors which might indicate what or how Malaysian consumers think of the E-wallet by measuring the effectiveness of E-Wallet in Malaysia. This study employed three independent variables which are Convenience, Security, and Cost-saving and was conducted to find out how the independent variables affect the online consumer behavior related to the E-Wallet system. This research can provide a comprehensive view on the consumers who are using E-Wallet so that E-Wallet developers can pay more attention to convenience, security, and cost saving factors to improve the effectiveness of E-Wallet system.

2 Literature Review

2.1 Consumer Purchase Decision Using E-Wallet

Rau [13] states that non-cash payments method has grown rapidly around the world. Physical cash is still widely used in Asia with only few countries advancing from cash faster than others. An example would be, Indonesia, with 99.5% of transactions are conducted using cash and the percentage rate of cashless transactions is growing annually at the rate of 23%. In Malaysia, 92.5% of transactions are in cash, but electronic transactions are growing at a comparatively modest 9% annually. Also, like other macro trends in the global economy, significant changes in purchase behavior occurs in developing countries, spurred by a strong surge of urbanization and a growing middle class. Based on the World Payments Report, about 40% of the major card schemes payment volumes now originate from emerging markets. With around 9.5% compound annual growth, Asia Pacific is still considered as the world's fastest growing cashless transactions market. Recently, China has recorded an increment of 32.7% annually in the number of non-cash transactions. Not only that, payment card market for Vietnam also has become one of the most fastest growing markets, increasing at an annual rate of 37% between 2008 and 2012.

The E-wallet is also part of this emerging technology that has been developing in recent years. Examples are Bitcoin in 2009, Samsung Pay in 2011, and Apple pay in 2014. However, e-commerce has been around since the 1980. The consumer market in Malaysia is still transitioning slowly as an emerging market and still under development, a majority of consumers with 29.6% under the age of 15, 65.4% between 15–64, and 5% remaining are above ages 65 stated by United Nations Statistics Division are still using physical cash. Issues of why the Malaysian population has not adopted this payment system could be that businesses/organizations within the country or region of Malaysia have not yet introduced a commercial widespread E-Wallet, the aged consumers with spending power have not shown any interest in such developments, difficulty in teaching the general public the uses of such a system, or perhaps there is reluctance from consumers in accepting a new concept of transaction.

2.2 Convenience of E-Wallet

In According to [3], E-wallet is a storage medium keeping financial and credit card information of the consumers which can be used to complete electronic or online transactions without having to re-enter the stored information at the time of the transaction. The uniqueness of E-wallet function allows consumers to have a hassle-free option to buy goods and services online. Lai and Ariffin [9] state that it provides a convenient and fast way of performing common online transactions. In recent times, there are many e-payment systems and one of them is PayPal.com. There is a different twist of e-payment programs offered by PayPal.com where they allow small businesses or consumers to transfer or send money using an email (Paypal.com [11]). For businesses, this service is convenient for them as cash can be transferred or sent quickly to their buyers or vendors. On the other hand, for consumers, parents with university or college students can easily transfer money into the student's account for books and tuition fees.

Many countries have already implemented the use of e-wallet as part of their daily purchasing transaction option for their consumers. E-wallet saves travelling time for consumers as all they need is just to scan their card or E-wallet app rather than taking their time looking for cash and queuing up to buy tickets. In relation to consumers in Malaysia, E-wallet provides a great convenience to consumers who purchase online frequently since E-wallet stores every financial data resulting in a fast hassle-free transaction. In addition, E-wallet reduces the line-up of users of public transportation with the tap 'n' go feature since the public transportation sector in Malaysia is growing rapidly nowadays.

2.3   Security

One of the main reasons why consumers and organisations try not to engage in e commerce activities is due to security. Since the development of ecommerce has grown rapidly, organisations and consumers are concerned about the arising security issues. The level of the security of the transaction is considered as a key issues and core to develop the e-commerce [18]. Fang et al. [7] state that customers want their personal information and identity to be kept confidential because they are afraid of their information being misused for online fraud. For both online and offline environment, trust is one of the important elements that need to be taken into consideration when making a transaction. The trust between customers and vendors is not easy to build on a public network such as the internet. Trust is used as a medium by consumers when they try to reduce the uncertainty of the transaction.

According to [7], lack of trust is often mentioned as the main reason for consumer not buying from online seller. Shaw [15] mentions that consumers are not fully protected from dishonest sellers as they might have hidden agenda just to get hold of consumer's information such as credit card number. The popularity of E-wallet can lead to the improvement in business transaction. Many researches have been done showing that E-wallet is interesting and consumers accept it as a transaction tool [2, 6, 14]. However, e-wallet faces some security issues. According to [15], due to anonymity and untraceable characteristics of it, E-wallet can be abused or stolen which result in losses to users. Many studies and surveys have been done regarding to E-wallet. Mobile e-wallet has many challenges. Some of the challenges are mentioned as follow [6]; (1) Hacking, (2) Security and transaction cost is expensive, (3) The risk of abusing and misusing E-wallet by the cause of anonymous and untraceable characteristics such as money laundry. However, researchers have made a lot of effort to control or eliminate the risks. For example, the introduction of E-Trading Laws, the legal reliability of E-signature, and password for offline e-money have been developed.

2.4 Cost Saving

E wallets are usually stored by consumers and can be used for almost all e commerce web sites. Furthermore, internet banking service and also provide almost similar features with e wallet such as transferring money to others, e-cash, e-pay and issuing e-checks. It is observed that more people are favourably accepting e-wallet in Malaysia as it helps an organization and government in many ways. One of the major thing E-wallet can benefit government and organization is moving from paper to electronic. Since electronic payments are cheaper than paper-based, and the speed of transactions is faster when shifted to electronic more people are now keen to use electronic payments. Paper comes in the form of checks, invoices from suppliers and payments to beneficiaries which create problem to the environment. Firstly, paper is inefficient and quite costly. Vohra (2015) has proven that issuing a paper check cost 90% more than issuing a direct deposit check. Electronic payments can help the government by removing unnecessary expenses. The high cost of processing paper checks is enough to beginning to move towards electronic payments. According to [3], an organization can provide greater transparency and avoid late-payment penalties by using e-payable. Moreover, by using e-payables, paper can be eliminated in the payment process through card-based payments.

E-wallet allows transferring money to smartphone with ease. For example, transferring money to family members in other countries all over the world which makes it an easier task for everyone without service charge and time of delay. According to [1], there are about 50 million consumers around the world that have accessed their accounts online, and 39 million of them have signed up for mobile banking. Mobile banking allows consumers to transfer funds, pay bills, deposit checks and do other transactions just by using smartphones. Chen [4] states that there are many banking tasks that can be conducted through online banking by bank customers such as loan application, investment purchase or sale and transaction such as utility bill payments. For consumers, E wallet can help them by showing the history of transaction made and keeping track of previous payments which set a limit for customers to spend more than they should.

Fig. 1 Demographic data of the respondents

	N	%
Gender		
Male	118	53.2
Female	104	46.8
Age Group		
18 – 25	181	81.5
26 – 35	20	9.0
36 – 45	6	2.7
Above 45	15	6.8
Income (RM)		
Below 1000	118	53.2
1001 - 3000	65	29.3
3001 - 5000	17	7.7
5001 - 10000	15	6.8
Above 10000	7	3.2
Ethnicity		
Malay	22	9.9
Chinese	130	58.6
Indian	33	14.9
Others	37	16.7
Education		
SPM/STPM	8	3.6
Diploma/Degree	172	77.5
Masters	15	6.6
PhD	2	0.9
Others	25	11.3

3 Research Methodology

The objective of this research is to identify the important factors affecting consumer's purchase decision using E-Wallet. Based on the literature, convenience, security, and cost saving factors were selected to measure the effectiveness of consumer's purchase decision using E-Wallet and research hypothesis were developed as follows.

H1 Convenience is positively related to Consumer Purchase Decision using E Wallet.
H2 Security is positively related to Consumer Purchase Decision using E Wallet.
H3 Cost-saving is positively related to Consumer Purchase Decision using E Wallet.

The survey questionnaires were distributed to 50 targeted respondents as a pilot test. In order to ensure each item is consistent, reliability test was conducted. There were 230 survey questionnaires distributed to the respondent and 222 survey responses were used for further analysis. The demographic data of the respondents is shown in Fig. 1.

A total of 222 respondents have participated in this study and according to Fig. 1 above, there are about the same number of males and females at 53.2% and 46.8% respectively. Furthermore, most of the respondents are from the age group of 18–25

Table 1 Result of reliability analysis

Variables	Retained items	CA value	Strength of CA
Independent variables			
Convenience	Q1–Q5	0.959	Very strong
Cost saving	Q1–Q6	0.743	Relatively strong
Security	Q1–Q5	0.767	Relatively strong
Dependent variable			
Consumer purchase decision using E wallet	Q1–Q8	0.743	Relatively strong

which is at 81.5%, followed by 9.0% at the age of 26–35, 2.7% for the age of 36–45 and finally 6.8% above 45 years old. For monthly income category, most of the respondents is categorized under the first category which is below RM 1,000 at 53.2%, followed by monthly income of RM 1,001–RM 3,000 at 29.3%, 7.7% at RM 3,001–RM 5,000, 6.8% at RM 5,001–RM 10,000 and 3.2% for monthly income of RM 10,000 and above. For ethnicity, the most significant percentage is Chinese at 58.6%, followed by Others at 16.7%, Indian at 14.9%, and finally Malay at 9.9%. In terms of education level, most of the respondents are from the undergraduate at 77.5%, followed by Others with 11.3%, Masters at 6.6%, SPM/STPM at 3.6% and 0.9% for PhD.

The reliability test is conducted for actual data collection and pilot test. After pilot test was conducted, some changes were made before the actual data collection. Cronbach's Alpha (CA) values are turned out to be higher than 0.6, the minimum acceptable value. Table 1 shows the result of the reliability test for each variable. The variables that have Cronbach's Alpha values more than 0.6 were kept for further analysis as stated by Flynn et al. (1994). Therefore, items that need to be further analysed are as follows; 8 items for Consumer Purchase Decision using E-wallet (0.743), 5 items for Convenience (0.959), 6 items for Cost Saving (0.743) and 5 items for Security (0.767).

4 Data Analysis

The model summary table provides information about the regression line's ability to account for the total variation in the dependent variable. As shown in Table 2, the R square value of 0.463 indicates that all the independent variables used in this research were able to explain 46.3% of the variation of the dependent variable of this research model, which could be interpreted as an acceptable fit of the research model to the data as shown in Table 3.

Based on the Table 4 which shows coefficients and p-values, all the predictors (Convenience, Security and Cost Saving) are positive where the significant level are

Table 2 Model summary

Models	R	R Square	Adjusted R square	Std. error of the estimate
1	0.681[a]	0.463	0.456	0.26894

[a]Predictors: (Constant), SEC, CS, CON
[b]Dependent Variable: CBE

Table 3 ANOVA

Model		Sum of squares	df	Mean square	F	Sig.
1	Regression	13.600	3	4.533	62.676	0.000[b]
	Residual	15.767	218	0.072		
	Total	29.367	221			

[a]Dependent Variable: CBE
[b]Predictors: (Constant), SEC, CS, CON

Table 4 Coefficients and P-values

Model		Unstandardized coefficients		Standardized coefficients	t	Sig.
		B	Std. error	Beta		
1	(Constant)	0.410	0.220		1.863	0.064
	CON	0.266	0.036	0.447	7.403	0.000
	CS	0.199	0.053	0.207	3.753	0.000
	SEC	0.151	0.049	0.184	3.057	0.003

[a]Dependent Variable: CBE

Table 5 Hypothesis testing

Hypothesis	Relationship	t-statistics	p-value	Result
H1	Convenience → Consumer Purchase Decision	7.403	0.000	Supported
H2	Security → Consumer Purchase Decision	3.057	0.003	Supported
H3	Cost saving → Consumer Purchase Decision	3.753	0.000	Supported

less than 0.05. This demonstrates that independent variables are making positive impacts on consumer purchase decision.

As shown in Table 4, the first hypothesis for relationship between convenience and consumer purchase decision using E-wallet was examined and turned out to be significant as the value of p-value is 0.000. Since the p-value of H1 is 0.000 it indicates that the hypothesis H1 is supported. It was concluded that convenience has a significant positive relationship with consumer purchase decision using E-wallet as shown in Table 5.

Although all the independent variables (convenience, security and cost saving) are positively related to the dependent variable, consumer purchase decision using E-wallet, convenience has the highest value of correlation coefficient of 0.624. This is also supported by Cheok et al. [5] that convenience seems to be the most important factor of using E-wallet, and it really affects the purchase decision of the consumer. In modern society, convenience has become one of the most important issues that consumers are facing. It was observed that the consumers tend to buy products and services online as it provides convenience and online banking is often used as the payment method (Wadhera et al. [19]). Furthermore, another benefit of using E-wallet is that customers do not have to keep entering their personal details during the payment process as all their personal detail such as bank account number, contact number and address are stored in the database. Payment methods have various levels of convenience.

According to Wadhera [19], for sellers, E-wallet system can provide convenient to them because all the payment made by the customers will be collected and directly transferred to the sellers' bank account via electronic transmission. Moreover, by using E-wallet system, sellers will get to see and manage their summary information and account detail. As mentioned above, [17] also state that E-wallet allows consumers to keep track of shipping and billing information with just a click at the sellers' sites or any other online platforms. Not only that, for multiple cards that use E-wallet system, they are providing the service that can store e-cash, e-checks and the consumers' credit card information. The second hypothesis for relationship between security and consumer purchase decision using E-wallet is examined and turned out to be significant as the value of p-value is 0.003. Since the p-value of H2 is 0.003 which fits the criteria of having $p < 0.01$, it indicates that the hypothesis H2 is supported. It was concluded that security has a significant positive relationship with consumer purchase decision using E-wallet. Based on the data analysis, security is also strongly related to consumer purchase decision using E-wallet with a correlation coefficient of 0.499. It shows that security is a good contributor to consumer purchase decision using which will increase the decision of payment method using E-wallet.

Nowadays, customers are more tech savvy and keen to enjoy the rapid advancement of the technology. It was revealed that security has become a significant factor that affects the consumer purchase decision using E-wallet. According to [8], both of the components of software and information can be available in E-wallet system. For the software component, it provides the security and safety with the encryption for the online transaction using E-wallet. Sudarno [16] states that E-wallet is not similar to credit card in terms of security. E-wallet helps consumers to conceal their private and banking information. Furthermore, the consumer's funds can only be accessed by E-wallet when the consumers make it available. This shows that when consumers make any payment, E-wallet does not access any other banking information and also does not log metadata which can be obtained by the third party [12]. Moreover, E-wallet provides a much better security by blocking hackers or third parties from stealing all the personal information that can be used for identity theft due to E-wallet acts as a proxy [20]. There are still concerns that online transaction is not safe compared to traditional payment. For example, the detail of customers' credit card can be easily stolen by hackers or the third party. Therefore, most of the E-wallet companies are trying to resolve this ongoing issue on security by providing extra layer of security which provide more secure and safe system in online transaction.

Uddin and Akhi [17] mention that most of the providers are also making use of the modern encryption technology in order to enhance E-wallet security. The third hypothesis for relationship between cost-saving and consumer purchase decision using E-wallet was examined and turned out to be significant as the value of p-value is 0.000. Since the p-value of H1 is 0.000, it indicates that the hypothesis H3 is also supported. It was concluded that cost saving has a significant positive relationship with consumer purchase decision using E-wallet. Based on the data analysis, it was noticed that there is a significant positive correlation between cost saving and consumer purchase decision. Cost-saving is also considered as a good contributor to consumer purchase decision which promotes the decision of using E-wallet as a payment method. E-wallet allows consumers to carry out any transactions just by using smartphones which not only saves cost but also saves a lot of the consumer's time. Chen [4] states that there are many banking related tasks that can be done through E-wallet by consumers such as transferring funds, pay utility bills with ease and efficiency. For retailers, implementing the E-wallet system is considered as an efficient and affordable alternative to track and receive payment from their customers. Most of the retailors are usually constrained with their budget and by using E-wallet system they can experience substantial cost saving. Furthermore, retailors also would not be charged for any upgrade in the E-wallet system compare to the point of sale (POS) system.

5   Conclusion

This research was conducted to examine the effectiveness of the E-wallet in Malaysia. Based on the analysis of the survey, convenience, cost saving and security were identified to make positive influence towards consumer purchase behavior using E-wallet. Although there is a vast potential of E-wallet in Malaysia, the adoption rate of this system is still low. This research can help the E-wallet providers and marketers to understand the important features of E-wallet so that they can improve the service quality of E-Wallet system.

The findings of this research can be used as a guide for E-wallet service providers to develop appropriate strategies to enhance the E-wallet services. Convenience, security and cost saving factors are appeared to be significant factors which might deserve extra attention from the software developers, online payment providers and also banking institutions when design of E-Wallet is in consideration.

Due to the fact that most of the respondents are university students, the responses may not be effectively reflecting the whole population in Malaysia. Another limitation that can be found throughout the process of the research is that most of the questionnaires were distributed through online. Even though it eased the data collection process, it would be desirable to conduct an in-depth interview for selected target groups as a follow up study of the survey questionnaire. An in-depth interview with focused group who have experience of using E-Wallet can result in more valid and comprehensive research findings.

References

1. Akhi, A.Y.: E-wallet system for Bangladesh an electronic payment system. Int. J. Modeling Optim. **4**, 217–219 (2014)
2. Basu, A., Muylle, S.: Assessing and enhancing e-business processes. Electron. Commer. Res. Appl. **10**(4), 437–499 (2011)
3. Chauhan, P.: E-wallet: The trusted partner in our pocket. Int. J. Res. Manag. Pharm. **2**(4) (2013)
4. Chen, C.: Perceived risk, usage frequency of mobile banking services. Manag. Serv. Qual. An Int. J. **23**(5), 410–436 (2013)
5. Cheok, L., Huiskamp, W., Malinowski, A.: E-commerce trends and payment challenges for online merchants: Beyond payment. Moduslink Whitepaper, pp. 1–25 (2014)
6. Eslami, Z., Talebi, M.: A new untraceable off-line electronic cash system. Electron. Commer. Res. Appl. **10**(1), 59–66 (2011)
7. Fang, Y., Qureshi, I., Sun, H., McCole, P., Ramsey, E., Lim, K.H.: Trust, satisfaction, and online repurchase intention: The moderating role of perceived effectiveness of E-commerce institutional mechanisms. MIS Q. **38**(2), 407–427 (2014)
8. Kemp, R.: Mobile payments: Current and emerging regulatory and contracting issues. Comput. Law Secur. Rev. **29**(2), 175–179 (2013)
9. Lai, P.C., Ariffin, Z.: Consumers' intention to use a single platform E-payment system: A study among malaysian internet and mobile banking users. J. Internet Bank. Commer. **20**, 3–4 (2015)
10. Niranjanamurthy, M., Dharmendra, C.: The study of E-commerce security issues and solutions. Int. J. Adv. Res. Comput. Commun. Eng. **2**(7) (2013)
11. Paypal.com.: Send money, pay online or set up a merchant account-paypal. Available from: https://www.paypal.com (2017). Last accessed 13 Jun 2017
12. Rathore, H.S.: Adoption of digital wallet by consumers. BVIMSR's J. Manag. Res. **8**(1), 69 (2016)
13. Rau, A.: E-payments in emerging markets. J. Paym. Strategy Syst. **7**(4), 337–343 (2014)
14. Ruiz-Martínez, A., Reverte, Ó.C., Gómez-Skarmeta, A.F.: Payment frameworks for the purchase of electronic products and services. Comput. Stand. Interfaces **34**(1), 80–92 (2012)
15. Shaw, N.: The mediating influence of trust in the adoption of the mobile wallet. J. Retail. Consum. Ser. **21**(4), 449–459 (2014)
16. Sudarno, B.E.P.: Analysis tracking online payment system. Int. J. Sci. Technol. Res. (IJSTR) **1**(2), 2 (2012)
17. Uddin, M.S., Akhi, A.Y.: E-wallet system for Bangladesh an electronic payment system. Int. J. Modeling Optim. **4**(3), 216 (2014)
18. Udo, G.J.: Privacy and security concerns as major barriers for E-commerce: A survey study. Inf. Manag. Comput. Secur. **9**(4), 165–174 (2001)
19. Wadhera, T., Dabas, R., Malhotra, P.: Adoption of M-wallet: A way ahead. Int. J. Eng. Manag. Res. (IJEMR) **7**(4), 1–7 (2017)
20. Westland, J.C.: Methods to assess the value of new technologies: The case of consumer sentiment towards digital wallet technology. Data-Enabled Discov. Appl. **1**(1), 2 (2017)

Designing of Domain Modeling for Mobile Applications Development

Haeng-Kon Kim

Abstract The term domain modeling has been considered as one of the important activities in systematic reuse. Is an activity that develops a generic model of a family of systems. It frames can be used to enable design layers, allowing the construction of complex structures and the reuse of development information. In this paper, we analyze the domain modeling support tool that retrieves objects from the candidate domain model to create frameworks from domain descriptions in a typical text format.

Keywords Domain modeling · Object identification · Design pattern
Program understanding · Components extraction

1 Introduction

The software test is the main factor for improving the quality of the software, for which it is necessary to generate different test cases according to certain coverage criteria, such as graphics, logic, input space and syntax [1].

As to domain engineering, it is a principle that refers to a family of similar systems to develop reusable components based on common aspects of the system [2]. The domain engineering process includes domain analysis, modeling, architecture, and implementation for reuse. In this process, domain analysis and modeling are human-intensive activities that require the extraction of domain model elements of information that is typically a text domain and a construction model [3]. Frames are sets of software that can increase efficiency in the development of complete applications in a particular area. Frameworks offer not only the reuse of software that implements the much-needed functionality, but also the reuse of design, standard cooperative structures that are typically used in applications in this field. Object-oriented technology to begin with is the most important technology currently used to use and develop frames. We have developed a domain modeling support tool that

H.-K. Kim (✉)
School of Information Technology, Daegu Catholic University, Gyeongsan, South Korea
e-mail: hangkon@cu.ac.kr

© Springer International Publishing AG, part of Springer Nature 2019   71
R. Lee (ed.), *Big Data, Cloud Computing, Data Science & Engineering*, Studies
in Computational Intelligence 786, https://doi.org/10.1007/978-3-319-96803-2_6

extracts the candidate domain model object. Extracted domain candidate objects could help domain engineers provide the ability to immediately evaluate domain descriptions and validation sources to compare domain engineering model objects. It can also be used to create a domain framework. It is based on rules of semantic extraction and cartography.

2   Domain Modeling Method, Environment and Frameworks

In recent years, as the flexibility of the platform and the reuse of software have been emphasized, a service-oriented architecture (SOA) has been developed. SOA systems now offer various services to potential users around the world, and the IT infrastructures of many companies have been commonly implemented as SOA-based systems [4]. Since the object-oriented software development model is considered to be more conducive to evolution and change, the domain modeling approach takes an object-oriented perspective. The goal is to apply object-oriented concepts and extend them to application domains. The domain modeling method is similar to other object-oriented methods when it is used to analyze and model a single system. Its novelty, and where it differs from other methods, is the way object-oriented methods are extended to model families of systems. Therefore, the method allows the explicit modeling of similarities and variations in a family of systems [5]. In a domain model, an application domain is represented by several views, so that each view presents a different perspective in the application domain. Four of the views, the aggregation hierarchy, the object communication diagram view, the generalization/specialization hierarchy, and the state transition diagram view have similar equivalents in other object-oriented methods used to model individual systems. On the status of domain modeling method, the aggregation hierarchy is used to model the optional object types, but not too often used by all members of the system family. In addition, the generalization/specialization hierarchy is also used to model variants of an object type, which are used by different members of the system family. The fifth view, the function/object dependency view, is used to explicitly model the captured variations in the domain model; each feature (optional domain requirement) is associated with the optional object types and the variant needed to support it. This provides the basis for defining which destination systems can be generated from the domain model [6]. This study involves an environment that permits the generation of target systems from a domain model as the general domain modeling environment.

The environments are called:

- Application generators and;
- Software system generators.

These generators are usually very domain specific because they have the structure and code for the application domain built into them. They also provide a means of adapting the code to generate a particular target system, either via the parameterization or via a user program written in a specific language of the domain. The objective of this manuscript is to have a domain-independent modeling and environment method to support the specification of a family of systems. In a domain model, an application domain is represented by several views, so that each view presents a different perspective in the application domain. Four of the views, the aggregation hierarchy, the object communication diagram view, the generalization/specialization hierarchy, and the state transition diagram view have similar equivalents in other object-oriented methods used to model individual systems. In the proposed domain modeling method, the aggregation hierarchy is also used to model optional object types, which are used by some but not necessarily all members of the system family. In addition, the generalization/specialization hierarchy is also used to model variants of an object type, which are used by different members of the system family. The fifth view, the function/object dependency view, is used to explicitly model the captured variations in the domain model; each feature (optional domain requirement) is associated with the optional object types and the variant needed to support it. This provides the basis for defining which destination systems can be generated from the domain model [7].

3　Development of the Domain Model

3.1　Component Identification

The component identification stage takes as input the business concept model and the use case model from the requirements workflow. It assumes an application layering that includes a separation of system components and business components. Its goal is to identify an initial set of business interfaces for the business components and an initial set of system interfaces for the system components, and to pull these together into initial component architecture. The business type model is an intermediate artifact from which the initial business interfaces are formed. It is also used later, in the component specification stage, as the raw material for the development of interface information models [8].

Any existing components or other software assets need to be taken into account too, as well as any architecture patterns you plan to use. At this stage it's fairly broadbrush stuff, intended to set out the component and interface landscape for subsequent refinement. To provide automated support for domain modeling, the semantics of the domain description must be extracted and correlated with the domain model elements. The classification of the domain description with respect to the predefined semantic representational structures makes it possible to establish the semantics of the description of the domain according to the category to which it belongs. The establishment of semantic representation structures and classification rules for the informal description of domains are key elements of the approach. The process

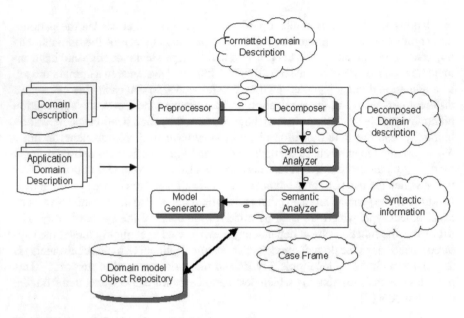

Fig. 1 Framework for domain model object generation

begins with the informal domain description text that is first preprocessed to format this text for use in other automated processes that follow. This pretreatment allows you to attach a unique identification number to each domain description sentence. The general framework is illustrated in Fig. 1.

Parsing is applied to establish the syntactic role of each word. This process provides syntactic models for use in semantic extraction. Case frames are used to represent the semantic information of domain descriptions. The structure of a case coffin is different for each primitive action. It is composed of an article in which it represents the types of activities and a body that represents the semantic objects of the concept. For different primitive actions, semantic objects are not the best to be different. The structure for each primitive action unites a different structure to describe its unique meaning. In the context of domain modeling, the words or expressions of collection objects are classified in the case bin category are used for the domain model objects concept of representing case frames is generally used for representing the thematic role of a word or phrase in a sentence. Each category has a fixed case structure. To illustrate, the following categorization is for action verbs: Existence: an action concerns the physical existence or the change of objects. In this category, the types of input, output, communication and recovery/storage function is included.

- Value: an action links the value of the object. The type of operation and the modification function are in this category.
- State: An action concerns the object mode. The types of control and mode functions are in this category.

- Evaluation: an action concerns the evaluation of the existence, value or state of the object for a given criterion. The identification and the decision are in this category.

3.2 Semantic Extraction

The concept of representing the case framework is usually very useful for representing the thematic role of a word or phrase in a sentence. Each category has a fixed frame structure.

- Existence: an action connects physical existence or a changing object. In this category, input, output, communication and recovery/storage types are included.
- Value: The action links the value of the object. The type of operation and the editing function are in this category.
- Status: The action is connected to the object mode. The types of functions and the control modes belong to this category.
- Classification: reports of actions, evaluating the existence, value or status of the object of a given criterion. Identification and touch belong to this category.

3.3 The Relation of Verbs Can be Classified in Two Field of Categories

1. Specialization: a relationship that represents the father and the child.
2. Aggregation: a relationship that represents.

```
M1: If (primitive action" identification) agent
     = common noun)
     = <agent> <object pattern>, <verb> action
M2: If (Primitive action= identification (object)
     = common noun)
M3: if (Primitive action= identification (domain! =Empty)
     =><domain> <object pattern>
M4: if (Primitive action) identification
(condition!)=Empty)= > <condition> condition
M5: if <object pattern>
 "noun phrase>< relation preposition><noun phrase>
| <noun>) => noun phrase> object, <noun> object,
<relational preposition> relation
Where,
<relation preposition>= in|of|on
M6: if ( Primitive action= Identification{ Object "Object) (do-
main)=object)=>relation belong to
```

Fig. 2 Classification of primitive action

```
Rid: 25
File Name: Domain Description
Action: Search
Agent: The system
Object: Candidates
Domain: designated repository of fingerprint information
Method:
Condition
```

Fig. 3 Classification of processing action

```
Rid:
File Name:
Action: <header pattern>
    Agent: <agent>
Object: <object>
Domain: <domain>
Method: <method>
Condition: <condition>
Where:
<identification header pattern> check|detect|discover|
recogncize|search
<Action> Identification header pattern
<Agent> Subject_Noun_Phrase <web pattern 1>
<Agent> Object_Noun_Phrase <web pattern 1>
<vern pattern> verb
<Domain> <Domain Pattern> Noun_Phrase
<Method> <method Pattern> Noun_Phrase
<condition> <condition Pattern> <Clause>
Condition pattern <Clause>
Condition pattern> if when|during|upon|unless|after|before|one
```

Fig. 4 Classification of final step action

Figure 2, describes a classification of primitive actions for the description of the expression domain [8] (Figs. 3, 4 and 5).

The parser has a submodule that is a syntax identifier. The syntax identifier identifies syntactic objects such as the subject, the object, and the predicate of the analyzed result by style. The identification process is performed according to the predefined grammar rules (Table 1).

The grammar analyzer has 2 subcomponents; semantic rule processor and case frame generator. The Semantic Rule Processor identifies words or phrases that meet semantic retrieval rules. This identification process uses lexical information and domain knowledge if it is available. The case framework builder uses the identified words or phrases to represent the sentence semantics of domain descriptions. The transformer has 2 subcomponents; Mapping Policy Processor and Template Object Organizer. The mapping rule processor identifies model objects based on the semantic

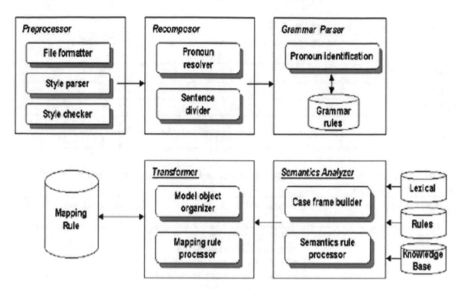

Fig. 5 Framework for domain model objects indentification

Table 1 Domain modeling elements

Domain model element	Domain model elements	Total number of model elements
Object	Controller, motor, water temperature, motor speed, flow, olive valve, fuel, 5 min, predefines value	15
Action	Activate, deactivate signal, regulate, operate	7
Relation	Controller of an oil, hot water home heating system, home temperature, combustion of furnace system, mim time in 5 min	5
Condition	Whenever the temp, falls below desired: once the speed is adequate temp reached is predefined value, after 5 min: within 5 s master switch off fuel flow shuts off	7

representation that is the case structure and the mapping rules. The identified model objects are organized by the model object organizer so that they can be understood by the domain description engineer (Table 2).

The second case is when the patterns and sentence structure are not treated with semantic extraction rules. In the text FLMS, the temperature of the house and the

Table 2 Domain model objects in FLMS

	Number of domain model elements	Number of model elements	Ratio (%)
Object	15	13	88
Action	7	6	89
Relation	5	4	80
Condition	7	7	100
Total	34	30	88

Table 3 Ratio between the domain model objects and system model elements in the generated candidate model objects

Domain model element	Domain model elements	Total number of model elements
Object	Electronics ten print submission, fingerprint, ten print card, criminal history file, electronic image, remote users, criminal history	22
Action	Scan, user, identify, search, send, submit, store, access, receive	9
Relation	Fingerprint belongs to electronic ten submissions includes live scan, ten print card print includes palm print, criminal history file has biological descriptor, file has personal identifier	11
Condition	When received by FBI	1

preset value is the second case and the operation and temperature of the house have a preset value are the first cases [9, 10] (Table 3).

4 Conclusion

A method has been developed to generate objects from the candidate domain model to provide automated support for modeling domain descriptions. The main feature of this method is to extract semantic information using case frames for the primitive actions in each domain description. The approach and architecture of the system developed to generate objects of the candidate conceptual model were presented. The main contribution of this research is the development of a methodological framework to provide automated support for the textual identification of domain model objects. The research has also contributed to the development of semantic extraction methods for the text of domain descriptions that include the use of case frames for domain descriptions, verbal and nominative classification structures, and dictionary information. The application of the decomposition method for the semantic extraction makes it possible to break down several descriptions of semantic domains into integrated semantic domain descriptions. This feature facilitates semantic extraction of domain descriptions from the text of informal domain descriptions. There are several areas for future research. A particular need that became evident in carrying out this research was the development of visualization and other forms of human computer interfaces for presentation of output formats. One possible application may be to convert the candidate model object lists to the input formats of the CASE tools that support the visualization of model objects. In this process, the normalization of the result would be necessary, including the elimination and reorganization of duplicate model objects. More research and experiments are needed to improve the accuracy of each module. In this task, developing a repository of common rules and domain knowledge could be important for supporting rules and reuse of knowledge.

Acknowledgements This Research was supported by the MSIP (Ministry of Science, ICT and Fu-ture Planning), Korea, under the ITRC (Information Technology Research Center) support program (IITP-2017-2013-0-0087) supervised by the IITP (Institute for Information and Communications Technology Promotion).

References

1. Saifan, A.A., Alsukhni, E., Alawneh, H., Sbaih, A.A.: Test case reduction using data mining technique. Int. J. Softw. Innov. (IJSI) **4**(4), 56–70 (2016)
2. Bjorner, D., Hlaváč, V., Jeffery, K.G., Wiedermann, J.: Domain engineering: a software engineering discipline in need of research. In: Wiedermann, J. (ed.) SOFSEM 2000: Theory and Practice of Informatics. Lecture Notes in Computer Science, vol 1963. Springer, Berlin (2000)
3. Kim, H.K.: Model driven engineering for crop monitoring applications. Int. J. Softw. Eng. Appl. **10**(6), 125–140 (2016)
4. Park, J., Seo, Y.-S., Baik, J.: A comparative analysis of reliability assessment methods for web-based software. Int. J. Softw. Innov. (IJSI) **1**(4), 31–44 (2013)
5. C. M. University: Domain analysis. http://www.sei.cmu.edu/domain-engineering/domain-anal.Html. Feb (1999)
6. Gomaa, H., Kerschberg, L.: Domain modeling for software reuse and evolution. http://mason.gmu.edu/~kersch/KBSE_folder/KBSEE_folder/RioCASEConference.html
7. Object oriented database systems. http://www.upriss.org.uk/db/42033lecture_one.html
8. Bjørner, D.: Domain engineering. In: Formal Methods: State of the Art and New Directions, pp. 1–41. Springer, London (2010)
9. Carnegie Mellon University: Domain analysis. http://www.sei.cmu.edu/domain-engineering/domain-anal.html
10. Prieto-Díaz, R.: Domain analysis: an introduction. ACM SIGSOFT Softw. Eng. Notes **15**(2), 47–54 (1990)

Performance Analysis of IoT Services Based on Clouds for Context Data Acquisition

Songai Xuan and Kim DoHyeun

Abstract Recently, Clouds are wildly used for huge data repository and Internet services in various fields. And IoT networks collect a context data and support the monitoring and control services using thing virtualization. We will build the connection between IoT and cloud, it is very useful, and supports intelligent services based on huge context data. This paper presents the comparison analysis of IoT services based on Clouds for huge context acquisition in large scale IoT networks. And, we develop AWS, Azure, and Google cloud based on IoT, and compare the IoT service of AWS, Azure, and Google Cloud by sending sensing data messages from IoT devices. The comparison helps users to choose easily IoT service based on Cloud. Hence, it is necessary to collect the context data easily and extract useful part for information analysis and usage in Cloud based on IoT.

Keywords IoT service based cloud · IoT network · Huge context data acquisition

1 Introduction

Internet of Things (IoT) infrastructures and systems have been deployed to various important area, frequently used for building smart environment, such as smart cities and smart homes [1]. Smart home is able to automatically sense the changes of home situations, dynamically response corresponding reactions and autonomously help its residents to make more comfortable lives [2]. For a smart home, there could be an IoT-based monitoring system using a tri-level context making model for context-aware services [3], there could be a device-level protections augmented with network-level security solutions to monitor network activity and detect suspicious behavior [4]. The smart environment makes people's lives faster and more convenient.

S. Xuan · K. DoHyeun (✉)
Department of Computer Engineering, Jeju National University, Jeju City, Republic of Korea
e-mail: kimdh@jejunu.ac.kr

S. Xuan
e-mail: xuansongai@foxmail.com

© Springer International Publishing AG, part of Springer Nature 2019
R. Lee (ed.), *Big Data, Cloud Computing, Data Science & Engineering*, Studies in Computational Intelligence 786, https://doi.org/10.1007/978-3-319-96803-2_7

IoT services has the constraint of IoT devices in a large scale networks. Because IoT devices has the limited computing resources, memory capacity, energy, and communication bandwidth. Many of these issues could be resolved by employing the Cloud-assisted Internet of Things as it offers large-scaled and on-demand net-worked computing resources to manage, store, process and share huge IoT data. It is an issue that how to deal with the large amount of information generated by the intelligent environment. Cloud computing is a good choice. Many researchers have already presented some survey of cloud computing, analyzed the key concepts and architecture [5] or introduced the cloud services of the IT companies [6]. Some researchers analyzed the authenticator-based data integrity verification techniques on cloud and IoT data [7]. The paper [8] presents an approach to the development of Smart Home applications by integrating Internet of Things (IoT) with Web ser-vices and Cloud computing, their approach focuses on Arduino platform, Zigbee technology, JSON data format, and cloud services.

The work presented in [9] presents a novel multilayered vehicular data cloud platform including an intelligent parking cloud service and a vehicular data mining cloud service by using cloud computing and IoT technologies. Some researchers developed a systematic comparator of the performance and cost of cloud providers called "CloudCmp" [10], which is able to help customers pick a cloud that fits their needs. In order to facilitate the users to choose a cloud easily, we present the comparison between the IoT cloud services of AWS, Azure, and Google Cloud based on sending sensing data messages from IoT devices.

Figure 1 shows the conceptual model of IoT services based on Clouds, then we use AWS, Azure, and Google Cloud based on IoT networks. The IoT devices are exactly the same, each of them send message to the IoT cloud (AWS IoT, Azure IoT Hub, and Google Cloud IoT Core). We compare and analyze the performance of three IoT cloud services based on the process and results of the experiment.

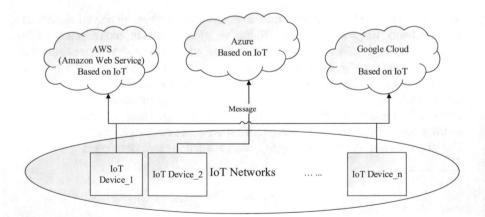

Fig. 1 Conceptual model of IoT services based on Clouds

There are many different IoT cloud platforms exist to meet the requirements of different user and application groups [11]. Some of them already have the service to support IoT devices connect the cloud (also called Cloud of Things) directly. Many other ones proposed their own Cloud of Things architecture [12], Cloud of Things framework [13], and Cloud of Things Middleware [14].

In this paper, we develop and compare the IoT services of AWS, Azure, and Google Cloud based on IoT devices. The comparison suppots to choose easily the IoT based cloud for huge context acquisition in large scale IoT network.

The remaining content of this paper is as follows. Section 2 describes the AWS IoT, Azure IoT Hub, and Google Cloud IoT Core, and IoT device. Section 3 presents development design of Cloud based on IoT. Section 4 shows experiment and comparison results of. Finally, Sect. 5 describes the conclusions and future.

2 Related Work

Amazon Web Services (AWS) [15] is a secure cloud services platform of Amazon.com, offering compute power, database storage, content delivery and other functionality to help businesses scale and grow. Explore how millions of customers are currently leveraging AWS cloud products and solutions to build sophisticated applications with increased flexibility, scalability and reliability. Figure 2 shows the AWS cloud layers.

Azure [16] is a comprehensive set of cloud services created by Microsoft; developers and IT professionals use to build, deploy, and manage applications through our global network of datacenters. Integrated tools, DevOps, and a marketplace offered supports efficient building of anything from simple mobile apps to internet-scale solutions. Figure 3 shows the main Azure cloud services.

Google Cloud [17], offered by Google, provides a set of management tools and a series of modular cloud services including computing, data storage, data analytics and machine learning. Figure 4 shows the Google Cloud services.

3 Development Design of Cloud Based on IoT

Figure 5 shows the detail architecture for the connection to AWS IoT. We used Raspberry Pi 3 Model B with Raspbian system as IoT device and used DHT11 Sensor (Temperature, Humidity). We installed AWS IoT Device SDK with python, and the device-to-cloud messages are send based on MQTT protocol.

Figure 6 shows the detail architecture for the connection to Azure IoT Hub. We used Raspberry Pi 3 Model B with Raspbian system as IoT device and used BME280 Sensor (Temperature, Humidity, and Pressure). We installed Azure IoT Hub SDK with JavaScript, and the device-to-cloud messages are sent based on MQTT protocol.

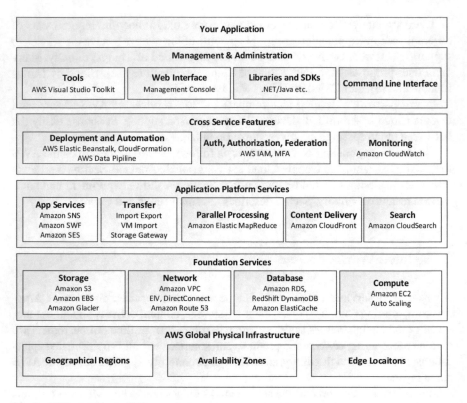

Fig. 2 AWS cloud layer [15]

Fig. 3 Main cloud service
of Azure [16]

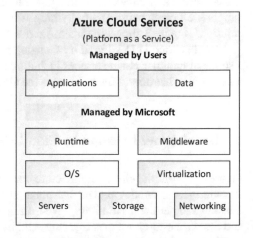

We use MQTT protocol for the publication of device-to-cloud messages. MQTT
[18] is a machine-to-machine (M2 M)/"Internet of Things" connectivity protocol.
It is a very lightweight publish/subscribe message transport. It is very useful to

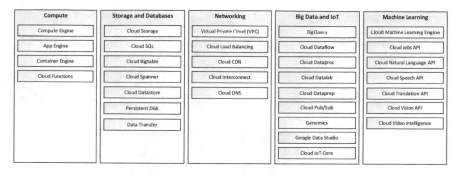

Compute	Storage and Databases	Networking	Big Data and IoT	Machine Learning
Compute Engine	Cloud Storage	Virtual Private Cloud (VPC)	BigQuery	Cloud Machine Learning Engine
App Engine	Cloud SQL	Cloud Load Balancing	Cloud Dataflow	Cloud Jobs API
Container Engine	Cloud Bigtable	Cloud CDN	Cloud Dataproc	Cloud Natural Language API
Cloud Functions	Cloud Spanner	Cloud Interconnect	Cloud Datalab	Cloud Speech API
	Cloud Datastore	Cloud DNS	Cloud Dataprep	Cloud Translation API
	Persistent Disk		Cloud Pub/Sub	Cloud Vision API
	Data Transfer		Genomics	Cloud Video Intelligence
			Google Data Studio	
			Cloud IoT Core	

Fig. 4 Google Cloud services [17]

Fig. 5 The configuration of connection between IoT device and AWS IoT

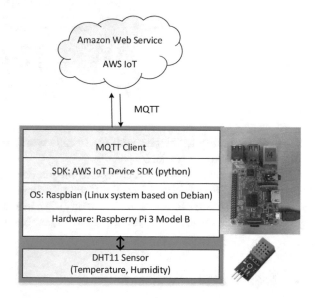

connect to remote locations requiring smaller code space and/or higher network bandwidth. For example, it has been used for sensors that communicate with brokers through satellite links, occasionally dial-up connections with health care providers, and home automation and small device scenarios. Because of its small size, low power consumption, minimized packets and efficient allocation of information to one or more receivers, it is also an ideal choice for mobile applications.

During the comparison, we use Raspberry Pies as the IoT devices to send the device-to-cloud messages to the IoT cloud services. The Raspberry Pi [19] is a powerful, small device that allows people of all ages to explore computing and learn how to use languages such as Scratch and Python to program. It can do everything necessary for desktop computers, from browsing the Internet and playing high-definition videos to making spreadsheets, word processing and playing games. Raspberry Pi

Fig. 6 The configuration of connection between IoT device and Azure IoT Hub

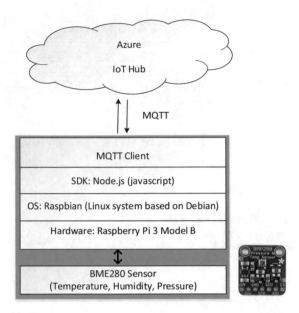

also has the ability to interact with the outside world and has been widely used in digital manufacturers projects.

Figure 7 shows the detail architecture for the connection to Google Cloud IoT Core. We used Raspberry Pi 3 Model B with Raspbian system as an IoT device and used DHT11 Sensor (Temperature, Humidity). We installed Google Cloud SDK with python, and the device-to-cloud messages are sent based on MQTT protocol.

4 Experiment and Results

Table 1 shows the simulation experiment environment. In each Raspberry Pi, we use Raspbian for IoT device and the cloud SDKs. For the SDK, we used AWS IoT Device SDK with python, Azure SDK with javascript, and Google Cloud SDK with python. All of the communication method is MQTT protocol, and we use Python 2 to compile the code and added DHT11 driver library and Paho MQTT Client library. And for our experiment, we considered the published message on AWS are able to showed more directly, the published message on Azure need to be download, and the published message on Google Cloud should show by the shell command.

Figure 8 shows the IoT device and sensors we used. We used Raspberry Pi 3 Model B with Raspbian system as IoT devices, which shows in Fig. 8a. We also used DHT11 Sensor (Temperature, Humidity) like Fig. 8b and BME280 Sensor (Temperature, Humidity, Pressure) like Fig. 8c.

Fig. 7 The configuration of
connection between IoT
device and Google Cloud

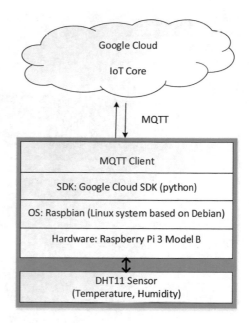

Table 1 Experiment environment

	AWS	Azure	Google Cloud
IoT device	Raspbian	Raspbian	Raspbian
SDK	AWS IoT Device SDK (python)	Node.js (javascript)	Google Cloud SDK (python)
Software	Python 2	Python 2	Python 2
Libraries	DHT11 driver, Paho MQTT Client	DHT11 driver, Paho MQTT Client	DHT11 driver, Paho MQTT Client
Webpage result	More directly	Download	Shell command

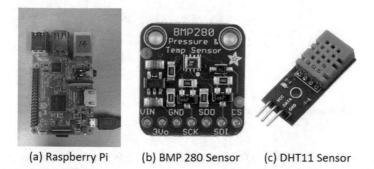

(a) Raspberry Pi   (b) BMP 280 Sensor   (c) DHT11 Sensor

Fig. 8 IoT device and sensors

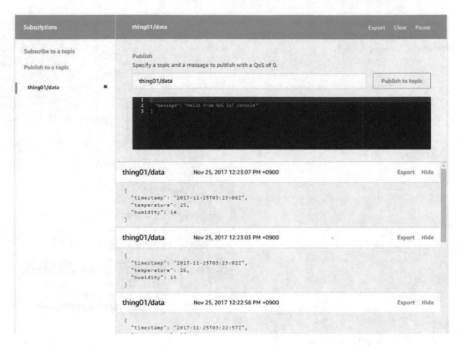

Fig. 9 Developed results in AWS

Fig. 10 Context messaged
published in AWS IoT topic

```
{
    "timestamp": "2017-11-25T03:23:06Z",
    "temperature": 25,
    "humidity": 14
}
```

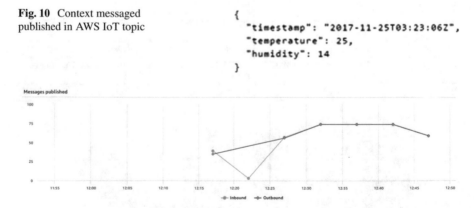

Fig. 11 The successful connections with AWS

Figure 9 shows the AWS IoT webpage, it's able to create a MQTT client and enter the same topic, which is able to show the published messages. Figure 10 shows one of the published messages, the format of this message is JSON.

Figure 11 shows the successful connections with AWS, When the connections become stable, the message publish speed is 0.25/s, the maximum speed is 0.25/s, the minimum speed is 0/s.

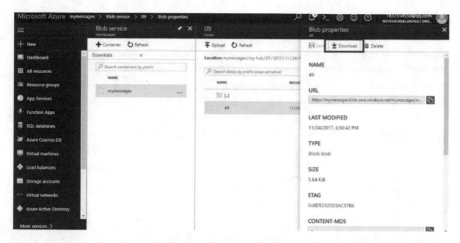

Fig. 12 Developed results in Azure IoT Hub

DC2messageIdH48571972-efad-4c68-a0ed-8c670ff41707SUBcorrelationIdH36d895b7-ca7e-4f4e-8a0c-6e5e458222b9$con
nectionDeviceIdCANraspberry-pi(connectionAuthMethod≈SOH{"scope":"device","type":"sas","issuer":"iothub","a
cceptingIpFilterRule":null}8connectionDeviceGenerationId$636465184990692747CANenqueuedTime82018-02-09T11:59
:46.1310000ZNULSTX~Time,2018-02-09-11:59:41,Temperature,24.16173,Pressure,807.6015SYNQžkI¬ÜÊ?K5áb_/Ý

Fig. 13 Context message download from Azure IoT Hub

Figure 12 shows the received message file in Azure, it's not able to show messages directly, we need to download the message file. In the figure, there is the Blob service of the storage we created, it is able to find the messages folder.

Figure 13 shows the downloaded message file from Azure, it is able to find the sensing data in the file. For example, as shown in the figure, there is the information include the message ID, the correlation ID, the device ID, Time (2018-02-09-11:59:41), the temperature value (24.16173), and the pressure value (807.6015).

Figure 14 shows the sum messages delivered to storage endpoints of Azure. When the connections become stable, the message publish speed is 0.55/s, the maximum speed is 0.55/s, the minimum speed is 0/s.

Figure 15 shows the received messages on Google Cloud Platform, we need google cloud shell command to check the message. And as shown in the figure, it is able to find the information include the message ID, the device ID, the device registry ID, the device registry location, the project ID, the temperature value (23.377046585083008), and the pressure value (983.5136108398438). (The detailed message shows in Fig. 16).

Figure 17 shows the publish message operations to the topic in Google Cloud Platform. When the connections become stable, the message publish speed is 0.81/s, the maximum speed is 0.95/s, the minimum speed is 0/s.

Table 2 shows the comparison result of IoT service based on Clouds in our experiments. For our last experiment (during 1 min), there are 15 messages published to

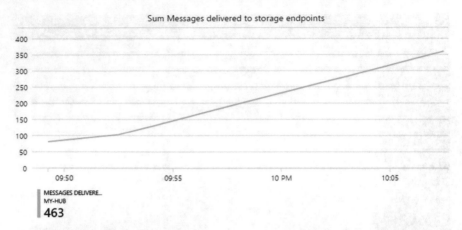

Fig. 14 Sum messages delivered to storage endpoints of Azure

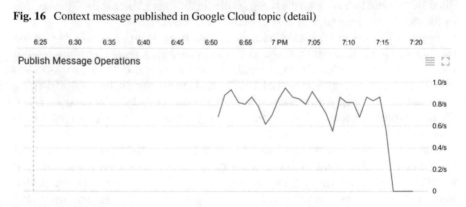

Fig. 15 Context message published in Google Cloud topic

{"data":[{"timestamp_Temperature":1518184480163,"Temperature":23.377046585083008},{"timestamp_Pressure":1518184480163,"Pressure":983.5136108398438}]}
UAYWLF1GSFE3GQhoUQ5PXiM_NSAoRRIJCBQFfH1wX1V1X1kaB1ENGXJ8Z3RtCOAEBUIFe1VYGQdoTml1JW8OGnt6aHVtWhoCBUNXdneDqY_s68FDZiU9XxJLLD5-MTdFQV4

Fig. 16 Context message published in Google Cloud topic (detail)

Fig. 17 The publish message operations to the topic in Google Cloud Platform

the topic of Cloud_A, 31 messages published to the topic of Cloud_B, and 50 messages published to the topic of Cloud_C. After many experiments, we considered the message publish speed of Cloud_A is 0.25/s, the message publish speed of Cloud_B is 0.55/s, the message publish speed of Cloud_C is 0.81/s.

Table 2 Comparison result of IoT service based on Clouds

	IoT service based on Cloud_A	IoT service based on Cloud_B	IoT service based on Cloud_C
Communication method	MQTT	MQTT	MQTT
Received messages (1 min)	15	31	50
Publish speed	0.25/s	0.55/s	0.81/s

Fig. 18 The mobile client of Google Cloud

Figure 18 shows the tested mobile client of Google Cloud, after click the "Get Message" button, the client is able to pull message from a Google Cloud subscription. For example, as shown in Fig. 18, the client pulled a message include the message ID (1524457045145), temperature value (28.0574893951426), and pressure value (973.6796264648438).

5   Conclusion

In this paper, we present the comparisons between the IoT services based on Clouds of AWS, Azure, and Google Cloud in IoT networks. We also compare and analyze the performance of three IoT services based on Clouds sending the sensing data messages between Clouds and IoT device. During the study, we gained more understanding about IoT services based on Clouds. In the future, we aim to make comparison by adding more IoT devices with more complicated comparison system.

Acknowledgements This research was supported by the MSIT (Ministry of Science and ICT), Korea, under the ITRC (Information Technology Research Center) support program (2014-1-00743) supervised by the IITP (Institute for Information and communications Technology Promotion), and this This work was supported by Institute for Information and communications Technology Promotion (IITP) grant funded by the Korea government (MSIT) (No.2017-0-00756, Development of interoperability and management technology of IoT system with heterogeneous ID mechanism). Any correspondence related to this paper should be addressed to DoHyeun Kim; kimdh@jejunu.ac.kr.

References

1. Botta, A., et al.: On the integration of cloud computing and internet of things. In: 2014 International Conference on Future Internet of Things and Cloud (FiCloud). IEEE (2014)
2. Chenishkian, S.: Building smart services for smart home. In: Proceedings of the IEEE 4th International Workshop on Network Appliances, pp. 215–224 (2002)
3. Kang, B., et al.: IoT-based monitoring system using tri-level context making model for smart home services. In: 2015 IEEE International Conference on Consumer Electronics (ICCE). IEEE (2015)
4. Sivaraman, V., et al.: Network-level security and privacy control for smart-home IoT devices. In: 2015 IEEE 11th International Conference on Wireless and Mobile Computing, Networking and Communications (WiMob). IEEE (2015)
5. Zhang, Q., Cheng, L., Boutaba, R.: Cloud computing: state-of-the-art and research challenges. J. Internet Serv. Appl. **1**(1), 7–18 (2010)
6. Zhou, M., et al.: Services in the cloud computing era: a survey. In: 2010 4th International Universal Communication Symposium (IUCS). IEEE (2010)
7. Liu, C., et al.: External integrity verification for outsourced big data in cloud and IoT: a big picture. Futur. Gener. Comput. Syst. **49**, 58–67 (2015)
8. Soliman, M., et al.: Smart home: integrating internet of things with web services and cloud computing. In: 2013 IEEE 5th International Conference on Cloud Computing Technology and Science (CloudCom), vol. 2. IEEE (2013)
9. He, W., Gongjun Y., Li Da, X.: Developing vehicular data cloud services in the IoT environment. IEEE Trans. Ind. Inf. **10**(2), 1587–1595 (2014)
10. Li, A., et al.: CloudCmp: comparing public cloud providers. In: Proceedings of the 10th ACM SIGCOMM Conference on Internet Measurement. ACM (2010)
11. Ray, P.P.: A survey of IoT cloud platforms. Futur. Comput. Inf. J. **1**(1–2), 35–46 (2016)
12. Abdelwahab, S., et al.: Cloud of things for sensing-as-a-service: architecture, algorithms, and use case. IEEE Internet Things J. **3**(6), 1099–1112 (2016)
13. Uehara, M.: A case study on developing cloud of things devices. In: 2015 Ninth International Conference on Complex, Intelligent, and Software Intensive Systems (CISIS). IEEE (2015)
14. Al-Jaroodi, J., et al.: CoTWare: a cloud of things middleware. In: 2017 IEEE 37th International Conference on Distributed Computing Systems Workshops (ICDCSW). IEEE (2017)
15. Amazon Web Service. aws.amazon.com
16. Microsoft Azure. azure.microsoft.com
17. Google Cloud Platform. cloud.google.com
18. MQTT. mqtt.org
19. Raspberry Pi. www.raspberrypi.org

Path-Based Integration Testing
of a Software Product Line

Jihyun Lee and Sunmyung Hwang

Abstract The testing of a product line is a more complex because variabilities spread across development processes and can be undetermined or absent while testing. Many existing studies of testing are focused on system testing, whereas integration testing is relatively rare. Because integration testing in SPLT tends to involve both domain testing and application testing, integration testing of SPL is necessary to clarify coverage problems in both testing stages. This is important in terms of thoroughness of testing but is also necessary to avoid redundant testing between two testing stages. In this paper, we propose the XX-MM-path-based integration testing method, which extends the MM-path-based testing method, and show how test coverage can be handled at both testing levels of domain and application testing. As a result, the MM-path-based integration testing method can be applied to the integration of common parts during domain testing without stub or driver implementation.

Keywords Software product line testing · Test coverage · Variability
Path-based testing · Integration testing

1 Introduction

In software product line engineering (SPLE), domain engineering sets up a common product line platform by identifying commonalities and variabilities, while application engineering develops individual products based on the platform. Each of the engineering processes of SPLE has a specific testing stage. The domain testing stage of the domain engineering process involves testing of common parts and produces reusable test artifacts known as domain test assets such as test plans, test cases, test

J. Lee (✉)
Department of Software Engineering, Chonbuk National University, Jeonju, Korea
e-mail: jihyun30@jbnu.ac.kr

S. Hwang (✉)
Department of Computer Engineering, Daejeon University, Daejeon, Republic of Korea
e-mail: sunhwang@dju.kr

© Springer International Publishing AG, part of Springer Nature 2019
R. Lee (ed.), *Big Data, Cloud Computing, Data Science & Engineering*, Studies
in Computational Intelligence 786, https://doi.org/10.1007/978-3-319-96803-2_8

data, and test scenarios. Meanwhile, the application testing stage of application engineering process has to achieve the efficient reuse of reusable test assets while it tests application-specific parts and performs regression testing of the parts of a product tested during domain testing [1].

The testing of a product line is a more critical activity than the testing of a single software product because the testing of a product line platform that will consist of multiple products is critical. In addition, as noted above, testing must reduce redundant test executions to achieve the benefits of a product line. Moreover, because variabilities spread across development processes and can be undetermined or absent while testing, software product line testing (SPLT) is highly complex. One type of SPLT study extends existing models such as activity diagrams and sequence diagrams to deal with variability, proposing methods to generate and run tests that are tailored to them [2–4]. Another type involves research on the introduction of a combinatorial interaction testing method to SPLT starting with work by Cohen et al. [5], focusing on sampling products that cover as many features as possible. Some studies, starting with that by Uzuncaova et al. [6], perform testing using individual products, but new products only rely on additional testing of the delta of the previous product. All of these studies correspond to system testing.

In terms of the test level, Neto et al. [7] considers that domain testing undertakes unit testing and integration testing, and the testing of products, i.e., system testing, is performed during the application testing stage. However, integration testing is described as being performed not only during domain testing but also during application testing. Unit testing of SPL can be done in the same manner as the testing of a single software product. There are two categories of SPL integration testing: one that extends the model to handle variability [8, 9], and another that performs integration testing based on the delta between product variants [8, 9]. In fact, there is not much research on integration testing compared to system testing. Ganesan et al. [10] proposes an incremental integration testing method that replaces and integrates driver and stub modules with actual modules after developing driver and stub modules when the actual variant values are not determined. This method is more concrete and realistic than previous methods. However, what these studies overlook is the coverage problem of SPL integration testing, which can only be performed during both domain testing and application testing.

Test coverage is the degree, expressed as a percentage, to which specified coverage items have been exercised by a test suite [11] and used as a device to measure the extent to which a set of test cases covers (or exercises) a program [12]. The coverage of integration testing in single software product testing represents the degree to which all interacting modules are integrated and the interaction is covered. MM-path-based testing defines coverage metrics for integration testing [12]. On the other hand, there is little discussion about test coverage compared to the system testing of SPL, and there is little discussion on this topic compared to single software product testing. Feature combination coverage and variant coverage, which are frequently used in existing SPLT studies, are types of coverage related to system testing. Thus far, coverage of integration testing has rarely been studied. Because integration testing tends to be performed during both domain testing and application testing, integration testing of SPL is required to clarify coverage problems at both testing levels. In this paper, we propose the XX-MM-path-based integration testing method, which extends the MM-path-based testing method, and show how test coverage can be handled at both testing levels of domain and application testing.

The paper is organized as follows: Sect. 2 describes SPL integration testing and the MM-path-based integration testing method used for single software products as background knowledge. Section 3 explains the concepts necessary to extend MM-path-based integration testing to SPLT and describes the XX-MM-path and possible coverage metrics. We conclude the paper in Sect. 4.

2   Related Work

This section describes the analysis results of existing integration testing approaches of SPL. Integration testing of SPL is relatively rare. In this section, integration testing studies of SPL can be explained by dividing them into model-based SPL integration testing and delta-oriented SPL integration testing.

2.1   Model-Based SPL Integration Testing

The methods in this category (1) explicitly model the variation points and variants using the model-based testing method of single software products [3, 8, 9], and (2) augment information such as test data to generate test cases from models [8] or use feature dependency diagrams to obtain interaction information [9]. All of these methods use domain architecture and product architecture and perform integration testing during both domain testing and application testing. The main difference between [3, 8] and [9] is that [3, 8] sees a component as the unit of integration, while [9] regards a feature as the unit of integration. Shi et al. [13] are similar to [9] in that this approach also considers features as interaction and integration units, but they differ in that [9] uses architecture and feature dependency diagrams while [13] uses a compositional symbolic execution approach to derive an interaction hierarchy.

2.2   Delta-Oriented SPL Integration Testing

The methods in this category analyze deltas between the architectures of product variants and perform test case prioritization based on these deltas [14, 15]. Lachman et al. [16] uses a delta-oriented architecture but improves the method based on the feature failure probability for test case prioritization. All of these methods are incremental SPL integration testing methods, introducing a test case prioritization method that occupies a large part of regression testing. However, unlike traditional regression testing, a test case with a higher priority means more interactions between components related to deltas. During risk-based test case prioritization, a test case with higher priority covers the interaction between components associated with a delta with a higher risk level.

In addition to the above two categories, two researchers [10, 17] propose incremental integration testing to generate test cases for integration testing from test cases for unit testing. When using this method, the variability is replaced by the stub or driver, and if the actual value of variability is determined, it is replaced by the actual module.

3   XX-MM-Path-Based Integration Testing of SPLs

The MM-path is a method-message path, which is a path that focuses on interaction with other units. In SPL, integration testing that focuses on interactions between units should also be performed. MM-path-based integration testing for the single software product testing can be applied in the same way for SPLT because the interactions in SPL eventually occur through method calls from other units. The main challenge in applying MM-path-based integration testing to SPLT is how to handle variability. Unit variability can be seen in unit testing. Therefore, integration testing only focuses on the interaction, so we only need to consider the variability of the methods involved in the interaction. There are several types of MM-paths in SPLT, depending on the variability. This section describes basic concepts, definitions and test coverage metrics of XX-MM-path-based integration testing for SPLs.

3.1   Basic Concepts and Definitions

In this section, we present the basic concepts and definitions necessary to define our method. Most definitions are tailored definitions for those of earlier work [12]. Based on the SPL characteristics, this study divides the statement fragment used in that study [12] as a basic concept for extracting the nodes constituting the path into the following two types:

- *Common statement fragment*: a sequence of statements or methods commonly used for developing member products of a product line.
- *Variable statement fragment*: a sequence of statements or methods used for some member products of a product line.

According to the above two concepts, the beginning and ending source and sink nodes of each MEP can be divided into the common and variable types, respectively:

- *Common source node*: a common statement fragment at which program execution begins or resumes. The first executable statement in a unit is clearly a common

Table 1 Types of MEPs with definitions

XX-MEPs	Definition
CC-MEP	Commonality/Commonality MEP starts with a common source node, passes only common statement fragments, and ends when it reaches a common sink node
CV-MEP	Commonality/Variability MEP starts with a common method or class, passes only common methods or classes that issue messages, and ends when it reaches a variable method or class that does not issue any messages of its own
VC-MEP	Variability/Commonality MEP starts with a variable method or class, passes only common methods or classes that issue messages, and ends when it reaches a common method or class that does not issue any messages of its own
VV-MEP	Variability/Variability MEP starts with a variable method or class, passes only common methods or classes that issue messages, and ends when it reaches a variable method or class that does not issue any messages of its own

source node. Source nodes also occur immediately after nodes that transfer control to other units or after variable source/sink nodes.

- *Variable source node*: a variable statement fragment at which program execution begins or resumes.
- *Common sink node*: a common statement fragment at which program execution terminates or transfers control to other units.
- *Variable sink node*: a variable statement fragment at which program execution terminates or transfers control to other units.

In SPLT, MEP is identical to single software product testing in that it is a series of paths from the source node to the sink node within the same unit. However, unlike in earlier work [12], the MEP in SPLT should be the basis of the coverage metric considering reuse while having common or variable nodes and minimizing redundancy. The MEP can be defined in various forms taking into account the combination of common or variable nodes. Therefore, in this study, the following two paths are defined as the MEP and MM-path consisting of the MEP.

Definition (XX-MEP): a sequence of nodes that begin with a common/variable source node and end with a common/variable sink node, with no intervening sink nodes.

Definition (XX-MM-path): a sequence of XX-MEPs linked by messages.

Based on these definitions, path-based integration testing for SPL classifies SPL-specific MEP types and MM-path types used for test coverage measurements. Path-based integration testing in SPL, in contrast to single software testing, requires the handling of variability. In other words, the test basis in the SPL can include variability; hence, the MEP should be defined while considering this. Table 1 describes the types of MEPs available.

The following is the result of the XX-MEP analysis of the NextDate product line tailored from the NextDate example in the aforementioned study [12], which consists

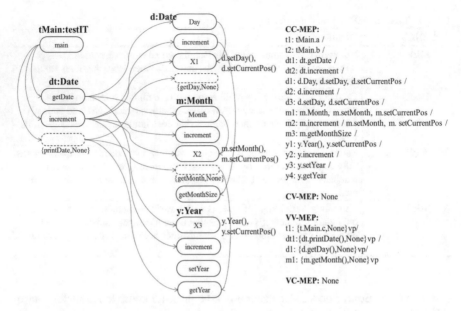

Fig. 1 Example of XX-MEP-paths for the NextDate product line

of five classes of mainId (testIt, Date, Day, Month, and Year). In Fig. 1, the dotted ellipse indicates a variable point, and the contents in '{}' on the right below the ellipse are variants.

In the case of XX-MM-path including CV-, VC-, and VV-MEP, which are MEPs containing variable slices, if the variation point is not bound, test data covering the path cannot be determined. It is not executable even if we can decide to test data. Whether to test these paths during domain testing or during application testing depends on the test strategy. For example, if we perform testing according to the common and reuse strategy, the test path containing variability is created during domain testing but the test execution is delayed until the binding time of application engineering. Table 2 describes the possible MM-path types according to the MEP types in Table 1.

Figure 2 shows an example of a CC-MM-path (solid arrow) and a VV-MM-path (dotted arrow) (the return path is omitted due to the complexity of the figure).

3.2 Test Coverage Metrics in XX-MM-Path

Existing methods overlook domain testing and reuse problems when dealing with test coverage problems in SPLT. However, the proposed method can distinguish between test coverage in domain testing and test coverage in application testing using XX-MEP. Possible SPL test coverage criteria in the proposed method are as follows:

Table 2 Types of XX-MM-paths with definitions

XX-MM-paths	Definition
CC-MM-path	Commonality/Commonality MM-path starts with a CC-MEP, passes only CC-MEPs, and ends when it reaches a CC-MEP of which the sink node sends no messages
CV-MM-path	Commonality/Variability MM-path starts with a CC-MEP, passes only CC-MEPs, and ends when it reaches a CV-MEP or VC-MEP of which the sink node sends no messages
VC-MM-path	Variability/Commonality MEP starts with a VC-MEP or CV-MEP, passes only CC-MEPs, and ends when it reaches a CC-MEP of which the sink node sends no messages
VV-MM-path	Variability/Variability MM-path starts with a XX-MEP, passes XX-MEPs, and ends when it reaches a XX-MEP of which the sink node sends no messages, where the combinations of MEPs are various but not included in the three types above

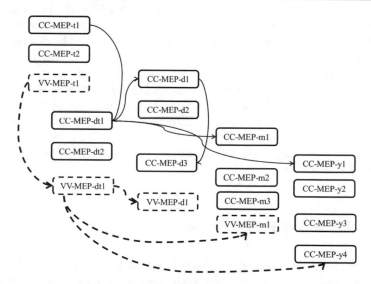

Fig. 2 XX-MM-path graph for the NextDate product line

- *CC-MM-path coverage*: This coverage metric includes only the CC-MM-path in the test run. This coverage has the advantage of being able to pre-verify commonality through a single product of a product line because it only designs and executes test cases that undergo commonalities.
- *CC- and CV-MM-path coverage*: This coverage metric is tested through CC-MM-path and CV-MM-path. Because CV-MM-path is the CV- or VC-MEP of the last MEP, to achieve this coverage during the domain test, a stub for the variability involved in the MEP should be developed. Stubs should be developed for every possible value that a variation point can have. In case variability is developed

during domain engineering, integration testing for variability is possible without stubs. Otherwise complete testing of actual variability is possible at integration testing of a member product.

- *CC- and VC-path coverage*: This coverage metric includes CC-MM-path and VC-MM-path in the test run. Given that VC-MM-path is the first MEP to be a CV- or VC-MEP, a driver must be developed for the variability involved in the MEP to achieve this coverage during domain testing. A driver must be developed for every possible value that a variation point can have.
- *All-path coverage*: All paths up to VV-MM-path are tested. Because the VV-MM-path includes several variants, to achieve this coverage in the domain test, a large number of stubs and drivers should be developed to replace the variability.

In SPLT, there may not be an executable software product during domain engineering. Therefore, the CC-MM-path can perform CC-MM-path testing after deriving a representative application. In addition, the VC-MM-path and the CV-MM-path contain variability at the beginning and end, implying that they can be tested during the domain test. Test coverage during application testing can be done by testing the VV-MM-path and application-specific parts that are not covered by domain testing.

The SPL test has a problem in that due to absent variants, numerous stub and driver implementations may be required during testing and, in the worst case, test execution may not be possible until the system is fully integrated. XX-MM-path-based integration testing not only distinguishes between common paths and variable paths but also enables variable paths to distinguish paths such as CV- or VC-MM-paths that can be tested before application testing. It is also possible to measure integration testing coverage both during domain testing and application testing, which was not possible before.

4 Conclusion and Future Work

In this paper, we describe a method for applying MM-path-based integration testing to SPLT that analyzes call-return paths between methods, performs integration based on paths, and defines test cases covering the paths.

As a result, the MM-path-based integration testing method can be applied to the integration of common parts during domain testing without any separate stub or driver implementation (or with minimal stubs or drivers). Next, MM-path-based testing addresses the problem that existing SPL integration testing did not address: coverage issues during domain testing and application testing. The coverage metrics defined in Sect. 3.2 can be used to determine the coverage criteria during domain testing. Application testing can test for application-specific parts while minimizing duplication testing according to the coverage values of domain testing. Finally, the proposed method makes it easy to describe variability explicitly in the path. Explicitly described variability affects test coverage decisions, but it is important input when deciding whether to cover variability in domain testing or in application testing. In domain testing, a separate stub or driver implementation is required, and decisions can be made based on the number of test codes required for these separate implementations.

On the other hand, the proposed method encounters a complexity problem when finding a path. Moreover, it is limited in that when using it, it is difficult to define a path which explicitly expresses the variability according to the variable distribution type and the variability implementation mechanism. Future studies should attempt to overcome these limitations and confirm the effectiveness of the proposed method.

Acknowledgements This research was supported by Basic Science Research Program through the National Research Foundation of Korea (NRF) funded by the Ministry of Education (2017R1D1A3B03028609).

References

1. Pohl, K., Böckle, G., van der Linden, F.: Software Product Line Engineering: Foundations, Principles, and Techniques. Springer (2005)
2. Reis, S., Metzger, A., Pohl, K.: A reuse technique for performance testing of software product lines. In: Proceedings of International Workshop on Software Product Line Testing (SPLiT) (2006)
3. Reuys, A., Reis, S., Kamsties, E., Pohl, K.: The ScenTED method for testing of software product lines. In: Käkölä, T., et al. (eds.) Software Product Lines: Research Issues in Engineering and Management, Chap. 13. Springer (2006)
4. Lamancha, B.P., Usaola, M.P., de Guzmán, I.G.R.: Model-driven testing in software product lines. In: Proceeding of the 25th IEEE International Conference on Software Maintenance (ICSM2009) (2009)
5. Cohen, M.B., Dwyer, M.B., Shi, J.: Constructing interaction test suites for highly-configurable systems in the presence of constraints: a greedy approach. IEEE Trans. Softw. Eng. **34**(5), 633–650 (2008)
6. Uzuncaova, E., Khurshid, S., Batory, D.: Incremental test generation for software product lines. IEEE Trans. Softw. Eng. **36**(3), 309–322 (2010)
7. Neto, P.A.M.S., Machado, I.C., McGregor, J.D., Almeida, E.S., Meira, S.R.L.: A systematic mapping study of software product lines testing. Inf. Softw. Technol. **53**(5), 407–423 (2011)
8. Reis, S., Metzger, A., Pohl, K.: Integration testing in software product line engineering: a model-based technique. In: Fundamental Approaches to Software Engineering, pp. 321–335 (2007)
9. Machado, I.D.C., Neto, P.A.M.S., Almeida, E.S.: Towards an integration testing approach for software product lines. In: IEEE IRI, pp. 616–623 (2012)
10. Ganesan, D., Lindvall, M., McComas, David, D., Bartholomew, M., Slegel, S., Medina, B.: Architecture-based unit testing of the flight software product line. In: Proceedings of the 14th International Software Product Line Conference (2010)
11. ISO/IEC/IEEE 29119-1:2013: Software and systems engineering–Software testing–Part 1: Concepts and definitions (2013)
12. Jorgensen, P.C.: Software testing, a craftmans's approach, 4th edn. CRC Press (2014)
13. Shi, J., Cohen, M., Dwyer, M.: Integration testing of software product lines using compositional symbolic execution. In: FASE 2012, pp. 270–284 (2012)
14. Lochau, M., Lity, S., Lachmann, R., Schaefer, I., Goltz, U.: Delta-oriented model-based integration testing of large-scale systems. J. Syst. Softw. **91**, 63–84 (2014)

15. Lachmann, R., Lity, S., Lischke, S., Beddig, S., Schulze, S., Schaefer, I.: Delta-oriented test case prioritization for integration testing of software product lines. In: Proceedings of the 19th International Conference on Software Product Line, pp. 81–90 (2015)
16. Lachmann, R., Beddig, S., Lity, S., Schulze, S., Schaefer, I.: Risk-based integration testing of software product lines. In: VaMoS 2017, pp. 1–8 (2017)
17. Li, J.J., Weiss, D.M., Slye, J.H.: Automatic system test generation from unit tests of exvantage product family. In: Proceedings of the International Workshop on Software Product Line Testing, pp. 73–80 (2007)

Automatic Generation of GUI Test Inputs Using User Configurations

Leegeun Ha, Sungwon Kang, Jihyun Lee and Younghun Han

Abstract GUI testing validates the functionality of a software-intensive system by exercising its GUI. Although much research on automatic generation of GUI test inputs has been conducted to reduce the cost of GUI testing, the current GUI test input generation techniques can miss testing the behavior of the system which is dependent on the user configuration, which may leave undetected the defects that appear only under a certain user configuration. In order to completely test the behavior of a system for all possible user configurations, this paper proposes a method that automatically generate GUI test inputs under all possible user configurations. Since testing all possible user configurations is infeasible for nontrivial systems, the method is designed such that the user can sample user configurations. Thus, the proposed method generates GUI test inputs for the behavior of the system dependent on user configurations in addition to the test inputs generated by the existing technique that does not consider user configurations. We implement our method as an automated tool for the Android framework and evaluate it with on five open-source Android apps. The evaluation results show that our method can indeed achieve additional code coverage while preserving code coverage achieved by the existing technique.

Keywords GUI testing · Test input generation · Software configuration

L. Ha (✉) · S. Kang · Y. Han
Korea Advanced Institute of Science and Technology, Daejeon, Republic of Korea
e-mail: betown@kaist.ac.kr

S. Kang
e-mail: sungwon.kang@kaist.ac.kr

Y. Han
e-mail: younghun.han@kaist.ac.kr

J. Lee
Chonbuk National University, Jeonju, Republic of Korea
e-mail: jihyun30@jbnu.ac.kr

© Springer International Publishing AG, part of Springer Nature 2019
R. Lee (ed.), *Big Data, Cloud Computing, Data Science & Engineering*, Studies in Computational Intelligence 786, https://doi.org/10.1007/978-3-319-96803-2_9

1 Introduction

Graphical user interface (GUI) testing is a way of assuring functional correctness of software-intensive systems through their GUI [1]. GUI is a popular mode of interaction with users, and more than 50% of source code implements the interface [2]. In GUIs, visual objects such as labels, buttons, menus, and scrolls, are displayed to a screen. A user provides inputs on the objects by clicking, typing, scrolling, etc., triggering change of the screen state [3]. GUI test cases consist of a sequence of events on the visual objects as inputs and comparison of the expected screen state and the given state [4].

Considering the nature of GUIs and their test cases, developing GUI test cases requires a great deal of effort; moreover, there may be undetected defects by executing with manually created test cases. Generating test inputs for GUI testing thus has gathered significant research attention from researchers [5] and is being actively researched with the rise of mobile apps.

Many modern software systems enable users to customize their functionalities and are designed to run under multiple user configurations. Current GUI test input generation techniques [6–9], however, fail to generate test inputs that can thoroughly explore the behavior of a system that varies depending on user configurations, with the consequence that defects that appear only under specific user configurations may not be revealed [10].

In this paper, we present a method to automatically generate GUI test inputs that can completely test the behavior that depends on user configurations. To achieve the goal, our method is designed as follows:

- Given the system under test (SUT), our method automatically generates GUI test inputs for all user configurations.
- Our method preserves test inputs generated by the existing technique [11] that our method utilizes and thus preserves code coverage of the existing technique that our method utilizes with respect to the behavior of the system independent of the user configuration.
- To integrate user configurations into test input generation step, our method extends existing model-based GUI test input generation techniques [11] and extracts (i) user configuration model, and (ii) event-configuration option traceability.
- Our method allows user configurations to be sampled using the configuration model, since any non-trivial system can have a large number of user configurations.

We implement our method as an automated tool based on GATOR [12], and evaluate our method on the Android framework, with respect to testing effectiveness and execution cost. The results from five open-source Android apps show that the proposed method improves line coverage of the existing technique. The remainder of the paper is structured as follows. Section 2 discusses related work. Section 3 shows the motivating example for our study, followed by descriptions of the proposed method in detail alongside an example application in Sect. 4. Section 5 presents our evaluation and analyzes the results. Finally, Sect. 6 concludes the paper and describes limitations.

2 Related Work

2.1 Automatic GUI Test Input Generation Techniques

GUI test data generation approaches often employ model-based approach in academia, where the model is often created semi-automatically or manually. Due to the model creation and maintenance cost, Memon et al. proposed the technique to reverse-engineer a model from SUT to generate test inputs [11].

Many of the recent research articles about automated GUI test input techniques have targeted the Android framework, where the developers adopt manual testing and lack automated test cases [12]. Choudhary et al. [13] classified Android GUI test input generation techniques into three strategies: (i) generating a stream of random test inputs, (ii) model-based exploration, and (iii) systematic exploration generating complex test inputs using symbolic execution, or evolutionary algorithms. While these techniques have their own heuristics to generate the inputs and each category still is actively researched [6–9], none of them have considered user configurations explicitly.

Moran et al. [14] proposed a hybrid approach of the random testing and the model-based testing for Android apps that incorporates the configurations of running operating system (e.g. options such as turning on Wi-Fi and GPS), during test execution and bug report generation. Still, our proposed method can be complementarily applied to the work to incorporate user configurations, which can be toggled by a sequence of user actions during runtime.

2.2 Testing Configurable Software Systems

Detecting configuration-specific defects requires testing under multiple configurations, which is inherently expensive, since the time to execute test cases increases not only with the cost of running test cases in a single configuration, but also with the number of configurations. In order to reduce the cost, studies have been conducted to sample the number of configurations to be tested [15, 16] and to select relevant test cases when configuration change is given [17]. Some of these works [16, 17] target systems without GUI, or assume a test suite or configuration space are provided [15, 17]. In contrast, our method differs since the method targets specifically GUI-based system and only implementation of SUT is required.

Jin et al. [18] suggested three requirements for practical configuration-aware testing/debugging techniques: (i) modeling the space of possible configurations (configuration space); (ii) relating configuration space to programmatic elements; and (iii) capturing accurate configuration when detecting a bug. We claim our method satisfies the requirement (ii), and satisfies partially (i), and (iii), with respect to user configurations accessible via GUI.

3 A Motivating Example

We illustrate our method with a simple, GUI-based Android app. Figure 1 shows a list of Activities[1] in the Android app, *Note*. The Activity on Fig. 1a is the main screen upon launch. A user is allowed to open the *Edit* Activity, or to delete a note in the screen. In the *Edit* Activity (Fig. 1b), the title and the content of a note can be modified with text input; the modification can be saved by clicking a button or discarded by the back event. Finally, the app allows a user to customize behavior of the app through the *Settings* Activity (Fig. 1c). There are two options on the screen: one can decide whether to confirm note deletion and the other can provide gesture controls to delete to edit a note.

Although the example app does not need complicated inputs to use all of its functionality, the resulting behavior may differ depending on the options enabled in the *Settings* Activity. For example, the example app may show a popup when "confirm delete" option is on. This implies potential effectiveness of test inputs may be enhanced if test inputs are executed under different user configurations. This finding is mentioned in the context of conventional software testing [10, 15] and also in the context of GUI testing [19]. However, considering user configuration with respect to automatic GUI test input generation has not received sufficient attention.

One approach to realize this is by replacing or modifying configuration files—commonly utilized mechanism for storing user configuration—before test execution. There are a couple of drawbacks: (i) some platforms that run the GUI framework (e.g. Android) may require root access to configuration files when testing outside of the SUT's instrumentation context, which hinders usability and applicability of the approach, and (ii) user configuration may be changed extraneously during test execution (e.g. by using preference menus) so that it is not guaranteed that the test suites are executed in the "specified" user configuration. The second drawback is more critical because some functionalities could not be tested under the overwritten configurations, and it is more difficult to track user configuration when encountering a bug.

Alternatively, user configuration can be modified through preference menus typically provided by GUI-based software. The challenge to implement this approach is in relating GUI inputs to configuration changes (i.e. knowing what event sequences modify which configuration options). Then, the relationship can be used to configure SUT through GUI and to avoid the events that change user configurations (e.g. accessing preference menus) when generating GUI test inputs. This can prevent overwriting user configurations during test execution given that the relationship is precisely established.

---

[1] Activity is one of the user interface components in Android app, which provides a single functionality to its user. It contains *Fragments* and *Widgets*.

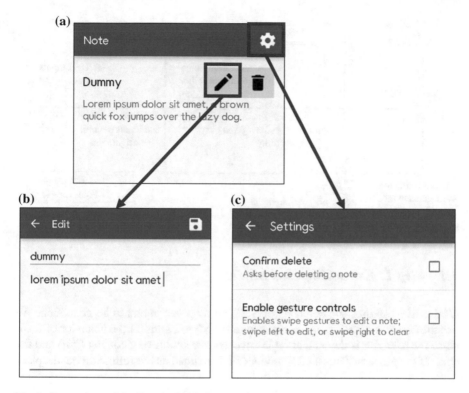

Fig. 1 Screenshots of the *Note* Android app

4 The Proposed Method

The proposed method is outlined in Fig. 2. It consists of four major steps: (1) GUI model extraction, (2) user configuration model extraction, (3) event-configuration option traceability extraction, and (4) generating test inputs for each user configuration. While the first step follows the general process of model-based GUI test input generation approaches, the other steps integrates user configurations into existing GUI test input generation approaches. Additionally, the method extracts the user configuration model from SUT, and the event-configuration option traceability from SUT and the previously extracted models. In the last step, all intermediate outputs are used to generate GUI test inputs under multiple user configurations. Extracting GUI model and user configuration model can be done in parallel, since they do not depend on each other. Subsequent subsections specify inputs or outputs of these steps. Then, we demonstrate application to the motivating example and describe how each step can be performed.

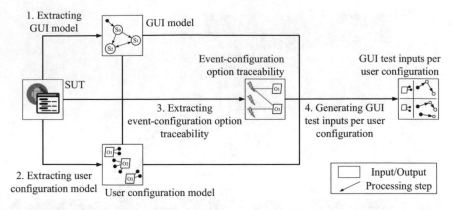

Fig. 2 Overview of the proposed method

4.1 Step 1: Extracting GUI Model

GUI model captures the behavior of SUT, guiding test inputs to be generated. We use the notation of the finite state machine (FSM) to represent the behavior of GUI, since the behavior is dependent on the events exercised (event context [20]) and the state of the process. Thus a GUI model GM is defined as the following quadruple:

$$GM = (S, E, s_0, T)$$

In GM, S denotes a set of states that can be differentiated by various comparison criteria [7]. E is a set of events that triggers transitions. s_0 ($s_0 \in S$) is an initial state. T ($T \subseteq S \times E \times S$) is a set of state transitions where each of them is labelled with an event. An event can have additional parameters, such as a text typed in case of type event of text boxes.

Figure 3 depicts a possible GUI model of the *Note* app. Each state is differentiated by Activity. Some events are omitted for brevity. Such GUI model can be extracted from SUT using static analysis on GUI layout files and source code, or "crawling" the SUT dynamically. The proposed method can be applied to either of them as the output is compatible with the GUI model we specified.

4.2 Step 2: Extracting Configuration Model

User configuration model encodes possible user configurations, and is represented by a set of configuration options the SUT has. Formally, we define user configuration model CM as a set of configuration options O_i, as below:

$$CM = \{O_1, O_2, O_3, \ldots\} \text{ where } O_i = \{v_1, v_2, v_3, \ldots\}$$

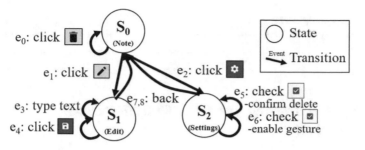

Fig. 3 An example GUI model for the *Note* app

	Configuration option	Assignable values
Table 1 User configuration model of the *Note* app	Confirm delete(O_{CD})	true, false
	Enable gesture(O_{EG})	true, false

Here, O_i is a range of values to which the option can be assigned. The range of values differs depending on the type the configuration option belongs to. For example, assuming a configuration option O_1 have Boolean value will model $O_1 = \{true, false\}$.

With CM, user configurations can be derived. Each configuration C_i is defined as n-tuple where n is the number of configuration options, and each element represents assignment of a value to the configuration option, as follows.

$$C_i = (O_{1,y1},\ O_{2,vx},\ \ldots)$$

In the *Note* app example, the user configuration model is extracted in Table 1. The *Note* app has two configuration options implemented with checkboxes (Fig. 1c), which can be turned on or off. This models each option Boolean type, having two values. Using the model, four user configurations can be derived.

The way to extract user configuration model differs across GUI frameworks and SUT. One is parsing resource files, when the GUI framework supports automatic generation of preference menus. Another is using static analysis by tracking API calls for accessing user configurations [21]. Note that extracting the accurate configuration model of non-trivial software is a daunting task. This is because user configuration can be accessed via several ways, such as API calls, preference menus in GUI, configuration files, and database [18]. However, we focus on preference menus in this study as they can change user configuration runtime through GUI inputs and would be handled for test input generation process (see Sect. 2).

Table 2 Event-configuration option traceability of the *Note* app

Event	Written configuration option
e_5	O_{CD}
e_6	O_{EG}

4.3 Step 3: Extracting Event-Configuration Option Traceability

Event-configuration option traceability is a key component in the proposed method. The traceability bridges GUI inputs and user configuration accesses, enabling automated GUI test input generation with controlled user configuration. The underlying rationale is that events are units in GUI test inputs and user configurations are determined by values of configuration options. Established from elements of extracted GM and CM, event-configuration option traceability EOT is defined as a relation between E and CM:

$$EOT = \{(e_1, \ O_1), (e_2, \ O_2), \ldots\}$$

In EOT, each element is a pair of an event ($e_i \in E$) and a written configuration option ($O_i \in CM$). While user configuration options can be either read or written during event execution, the traceability encodes the latter to detect user configuration change in GUI model and generate GUI test inputs per user configuration.

Table 2 lists event-written configuration option pairs extracted from the *Note* app. Since checking checkboxes in Fig. 1c causes corresponding configuration option change, the pair (e_5, O_{CD}) and (e_6, O_{EG}) are added to the event-configuration option traceability. Again, such traceability can be extracted by exploiting knowledge of GUI frameworks. In the case of modern GUI frameworks where resource files are used to generate preference menus, the events generated along preference GUI menus are traced to the configuration options extracted from the resource files. In the more complicated case where API calls are used to write configuration options, the events that execute API calls can be traced to the written configuration options.

4.4 Step 4: Generating Test Input for Each User Configuration

In the last step, GUI test inputs can be generated by using GUI model, user configuration model, event-configuration option traceability extracted. The output of the proposed method is a set of test inputs TI, whose elements are identified by a sequence of events ($i_i \in E^*$).

This step is comprised of three stages: (i) generating GUI test inputs, (ii) sampling the user configurations to be tested, (iii) generating extended GUI test inputs which execute GUI test inputs under the sampled user configurations. The executing cost and effectiveness of resulting GUI test inputs vary depending on how stage (i) and (ii) are performed.

4.4.1 Generating GUI Test Inputs

This stage generates non-user configuration writing event sequences using GM and EOT. The sequences can be obtained by walking the GUI from the initial state according to a traversal strategy (e.g. depth-first strategy) to satisfy test adequacy criteria [3] (e.g. event coverage, length-k event sequence coverage). In order to prevent user configuration change during test execution, the events which have traceability to user configurations are ignored. While this can miss some behaviors due to omitted events, we have observed the events related to setting user configurations tend to be isolated from the others. Thus it is assumed that it is possible to generate inputs to test the functionalities of the SUT aside from the "preference menus".

4.4.2 Sampling User Configurations

We sample user configurations to be tested from CM, as exhaustively testing all user configurations is infeasible for nontrivial systems. User configurations can be sampled using combinatorial approaches with a specified interaction strength [22], such as testing possible values of configuration options at least once (1-way) or n-way combinations of the options values at least once.

4.4.3 Generating GUI Test Inputs for Sampled User Configurations

Finally, the GUI input prefix, which alters the user configurations, is added before the previously generated test inputs, for each sampled user configuration. The prefix is the shortest path to the events in EOT (e.g. a sequence of events that accesses preference menu); these events in EOT are assigned with parameters whose values are in align with the user configuration to be tested. This ensures that the subsequent GUI test inputs are executed under the sampled user configurations. Given that there are m user configurations to be tested and n GUI test inputs are previously generated, the total number of resulting GUI test inputs is $(n+1) \times m$, since the previously generated test inputs are used across user configurations to preserve test effectiveness of those inputs.

Test input	Sequence of events
i_0	$e_0 \rightarrow e_1 \rightarrow e_3 \rightarrow e_4 \rightarrow e_7$
i_1	$e_2 \rightarrow e_8$

Table 3 GUI test inputs generated from Fig. 3

Table 4 GUI test inputs per user configuration

User configuration	Test inputs with GUI input prefix (underlined)
$(O_{CD,true}, O_{EG, true})$	$\underline{e_2 \to e_5 \text{ (true)} \to e_6 \text{ (true)}}$, $e_0 \to e_1 \to e_3 \to e_4 \to e_7$, $e_2 \to e_8$
$(O_{CD,false}, O_{EG, false})$	$\underline{e_2 \to e_5 \text{ (false)} \to e_6 \text{ (false)}}$, $e_0 \to e_1 \to e_3 \to e_4 \to e_7$, $e_2 \to e_8$

Table 3 shows examples of the GUI test inputs generated from the GUI model of the *Note* app (see Fig. 3). The GUI test inputs after applying remaining stages are listed in Table 4. The GUI inputs i_0, i_1 are obtained using the depth-first exploration strategy to achieve event coverage. Note that events e_5 and e_6, which are traced to configuration option O_{CD} and O_{EG} respectively, are omitted in Table 3. Instead, these events appear in the GUI input prefixes after e_2 (see Table 4) with a parameter (e.g. *true* to turn on the checkbox and vice versa), manipulating user configurations before i_0 and i_1 begin. Also, two of the four user configurations, i.e. $(O_{CD, true}, O_{EG, true})$ and $(O_{CD, false}, O_{EG, false})$, are used to test all the configuration option values at least once, based on the observation that many configuration options do not interact with each other [23]. As the result, six distinct GUI test inputs are generated so that two GUI test inputs can be executed under two user configurations.

5 Evaluation

5.1 Experimental Design and Setup

We evaluated the proposed method by comparing the effectiveness and execution cost of the GUI test inputs generated with and without user configuration. To that end, we implemented an automated tool for the Android framework and measured (i) line coverage, and (ii) the number of generated events after executing the generated inputs from (i) existing GUI model extraction technique [24] and test input generation strategy [25] (the existing technique), and (ii) the proposed method applied to the existing technique. Line coverage is measured for indicating potential test effectiveness using EMMA code coverage tool [26]. Besides, the number of events is counted to analyze how much test execution cost is increased when applying the proposed method. The experiment is conducted on a Windows 10 machine with a 3.4 GHz quad core processor, 8 GB RAM, which hosts Google Nexus 5X, Android 6.0 virtual device with x86 ABI; the machines are signaled using HTTP and Android Debug Bridge (ADB). The subjects are filtered by (i) applicability of static GUI model reverse-engineering tool [24], and (ii) existence of preference menus supported by the Android framework from the AndroTest benchmark suite [13] and the subjects in [24].

5.2 Tool Implementation

The tool uses an Android package file (.apk file) as input. GMs are statically obtained using the window transition graphs retrieved from the GATOR [27] toolkit. For extracting CMs and EOTs, the implemented tool parses `PreferenceScreen` resource files, which are used by the Android framework to construct preference menus (`PreferenceActivity`). The user configurations are sampled at the 1-way interaction strength, and the GUI test inputs are generated by deploying the depth-first strategy on the extracted GM, to achieve event coverage. To minimize the number of test cases that are not executable, which can be caused by imprecision in the static analysis we used, our tool generate test cases from event sequences from the initial state that are short. Lastly, the resulting test inputs are formatted to the UIAutomator test automation framework scripts. Note that we extended GATOR to support `PreferenceActivity`, the depth-first strategy, and the UIAutomator format.

5.3 Result

Table 5 shows the line coverage and the number of generated events for each subject. It shows that our method achieved 2.1% extra line coverage on average, including the coverage achieved by the existing technique. In order to mitigate internal validity, we examined the coverage reports and confirmed that the additionally achieved coverage stems from the configuration-dependent code, not from the repetition of test execution. In terms of the number of generated events, our method created 4.1% events to use preference menus per user configuration on average, in addition to the number of non-user configuration writing inputs multiplied by the number of sampled user configurations. The preference menus required more events for the subjects with more configuration options to be changed.

   This line coverage boost of the proposed method—from 0 to 4.8%—may seem small compared to the potential test execution cost. After manual investigation on the source code, intermediate artifacts, and test inputs from subjects and other open-source apps, the following causes are identified:

- The portion of programmatic elements (e.g. statements, methods) which is dependent on the user configuration may be limited; a similar trend was also observed in the related work [15, 17].
- Accessing some functionalities affected by configuration options requires test inputs aside from GUI events. For instance, subject *a2dp* has a configuration option

Table 5 Experiment result

Subject			Line coverage (%)		Number of generated events	
Name	Executable LOC	# of user configurations	Existing technique	Proposed method	Existing technique	Proposed method
a2dp	3430	7	33.1	35.2	1123	7882
autoanswer	386	6	8.8	8.8	7	48
dialer2	885	11	43.5	44.1	862	9515
lockpatterngenerator	576	7	79.0	82.1	122	875
MunchLife	161	3	79.0	80.6	93	288

GPS Listener Timeout, which customizes functional behavior of a background task that uses GPS service of the operating system.

- Some configuration options have complex input space such as character strings or constraints (e.g. an integer in specified range) to be configured with trivial values [21]. Automating configuration of such options will require in-depth analysis of source code.

6   Conclusion

The contributions of our research are as follows:

- We developed a method that generates test inputs for the behavior of the system dependent on user configurations, which is missed by the current GUI test input generation techniques, while not compromising code coverage of the underlying existing model-based technique.
- We implemented the method as a fully automated tool using the GATOR [24] Android static analysis toolkit.

Although our tool targeted Android for evaluation, we believe our method can be generalized for other GUI frameworks and model-based GUI test input generation techniques. For instance, if SUT is a Java desktop application on the Swing framework, our method can be implemented based on the GUITAR [28] tool.

We observed two limitations of our method for practical use. First, the proposed method entails increase in test execution cost compared to the increased code coverage. This is because the proposed method executes all the generated GUI inputs for each user configuration, although individual configuration options customize specific functionalities of SUT. The superfluous test inputs can be reduced as discussed in the work of Qu et al. [18]. Second, we selected a static model extraction technique and a single configuration access mechanism for implementation and evaluation, while real-world GUI-based software has highly dynamic GUI and uses various configuration access mechanisms [19]. We plan to extend our method to overcome these limitations.

Acknowledgements This research was supported by the MSIT (Ministry of Science and ICT), Korea, under the ITRC (Information Technology Research Center) support program (IITP-2018-2013-0-00717) supervised by the IITP (Institute for Information and communications Technology Promotion) and by Basic Science Research Program through the National Research Foundation of Korea (NRF) funded by the Ministry of Education (2017R1D1A3B03028609).

References

1. Ammann, P., Offutt, J.: Introduction to Software Testing, 1st edn. Cambridge University Press (2008)
2. Myers, B.A., Rosson, M.B.: Survey on user interface programming. In: Proceedings of the SIGCHI Conference on Human Factors in Computing Systems, pp. 195–202. ACM (1992)
3. Memon, A.M., Soffa, M.L., Pollack, M.E.: Coverage criteria for GUI testing. In: ACM SIG-SOFT Software Engineering Notes, vol. 26, Issue 5, pp. 256–267 (2001)
4. Memon, A.M., Pollack, M.E., Soffa, M.L.: Automated test oracles for GUIs. In: ACM SIG-SOFT Software Engineering Notes, vol. 6, pp. 30–39. ACM (2000)
5. Banerjee, I., Nguyen, B., Garousi, V., Memon, A.: Graphical user interface (GUI) testing: systematic mapping and repository. Inf. Softw. Technol. **55**(10), 1679–1694 (2013)
6. Zeng, X., Li, D., Zheng, W., Xia, F., Deng, Y., Lam, W., Yang, W., Xie, T.: Automated test input generation for Android: are we really there yet in an industrial case? In: Proceedings of the 2016 24th ACM SIGSOFT International Symposium on Foundations of Software Engineering, pp. 987–992. ACM (2016)
7. Baek, Y.-M., Bae, D.-H.: Automated model-based Android GUI testing using multi-level GUI comparison criteria. In: 2016 31st IEEE/ACM International Conference on Automated Software Engineering (ASE), pp. 238–249. IEEE (2016)
8. Su, T., Meng, G., Chen, Y., Wu, K., Yang, W., Yao, Y., Pu, G., Liu, Y., Su, Z.: Guided, stochastic model-based GUI testing of Android apps. In: Proceedings of the 2017 11th Joint Meeting on Foundations of Software Engineering, pp. 245–256. ACM (2017)
9. Mao, K., Harman, M., Jia, Y.: Sapienz: multi-objective automated testing for android applications. In: Proceedings of the 25th International Symposium on Software Testing and Analysis, pp. 94–105. ACM (2016)
10. Memon, A., Porter, A., Yilmaz, C., Nagarajan, A., Schmidt, D., Natarajan, B.: Skoll: distributed continuous quality assurance. In: Proceedings 26th International Conference on Software Engineering, 2004. ICSE 2004, pp. 459–468. IEEE (2004)
11. Memon, A.M, Banerjee, I., Nagarajan, A.: GUI ripping: reverse engineering of graphical user interfaces for testing. In: WCRE, p. 260 (2003)
12. Kochhar, P.S., Thung, F., Nagappan, N., Zimmermann, T., Lo, D.: Understanding the test automation culture of app developers. In: 2015 IEEE 8th International Conference on Software Testing, Verification and Validation (ICST), 2015, pp. 1–10. IEEE (2015)
13. Choudhary, S.R., Gorla, A., Orso, A.: Automated test input generation for android: are we there yet?(E). In: 2015 30th IEEE/ACM International Conference on Automated Software Engineering (ASE), pp. 429–440. IEEE (2015)
14. Moran, K., Linares-Vásquez, M., Bernal-Cárdenas, C., Vendome, C., Poshyvanyk, D.: Automatically discovering, reporting and reproducing android application crashes. In: ICST 2016 (2016)
15. Qu, X., Cohen, M.B., Rothermel, G.: Configuration-aware regression testing: an empirical study of sampling and prioritization. In: Proceedings of the 2008 International Symposium on Software Testing and Analysis, pp. 75–86. ACM (2008)

16. Song, C., Porter, A., Foster, J.S.: iTree: efficiently discovering high-coverage configurations using interaction trees. IEEE Trans. Softw. Eng. **40**(3), 251–265 (2014)
17. Qu, X., Acharya, M., Robinson, B.: Impact analysis of configuration changes for test case selection. In: 2011 IEEE 22nd International Symposium on Software Reliability Engineering (ISSRE), pp. 140–149. IEEE (2011)
18. Jin, D., Qu, X., Cohen, M.B., Robinson, B.: Configurations everywhere: implications for testing and debugging in practice. In: Companion Proceedings of the 36th International Conference on Software Engineering, pp. 215–224. ACM (2014)
19. Bae, G., Rothermel, G., Bae, D.-H.: Comparing model-based and dynamic event-extraction based GUI testing techniques: an empirical study. J. Syst. Softw. **97**, 15–46 (2014)
20. Yuan, X., Cohen, M.B., Memon, A.M.: GUI interaction testing: Incorporating event context. IEEE Trans. Softw. Eng. **37**(4), 559–574 (2011)
21. Rabkin, A., Katz, R.: Static extraction of program configuration options. In: Proceedings of the 33rd International Conference on Software Engineering, pp. 131–140. ACM (2011)
22. Cohen, M.B., Snyder, J., Rothermel, G.: Testing across configurations: implications for combinatorial testing. In: ACM SIGSOFT Software Engineering Notes, vol. 31, Issue 6, pp. 1–9 (2006)
23. Reisner, E., Song, C., Ma, K.-K., Foster, J.S., Porter, A.: Using symbolic evaluation to understand behavior in configurable software systems. In: Proceedings of the 32nd ACM/IEEE International Conference on Software Engineering, vol. 1, pp. 445–454. ACM (2010)
24. Yang, S., Zhang, H., Wu, H., Wang, Y., Yan, D., Rountev, A.: Static window transition graphs for android (T). In: 2015 30th IEEE/ACM International Conference on Automated Software Engineering (ASE), pp. 658–668. IEEE (2015)
25. Amalfitano, D., Fasolino, A.R., Tramontana, P., De Carmine, S., Imparato, G.: A toolset for GUI testing of Android applications. In: 2012 28th IEEE International Conference on Software Maintenance (ICSM), pp. 650–653. IEEE (2012)
26. EMMA: a free Java code coverage tool. http://emma.sourceforge.net/
27. GATOR: Program Analysis Toolkit For Android. http://web.cse.ohio-state.edu/presto/software/gator/
28. Nguyen, B.N., Robbins, B., Banerjee, I., Memon, A.: GUITAR: an innovative tool for automated testing of GUI-driven software. Autom. Softw. Eng. **21**(1), 65–105 (2014)

Execution Environment for Process Defined in EPF

Jeong Ah Kim

Abstract To analyze the processes applied to the existing R&D projects and standardize the analytical results into the RD concept, we defined a framework based on SPEM, which further defines method class, method component, and process component as framework components. Based on these definitions, we established a model that can create a delivery process. We further constructed an environment that can be monitored while executing the delivery process.

Keywords Process execution · SPEM · Process modeling notation
Process execution

1 Introduction

Process monitoring is related with process visibility so that every stakeholder can get insight into the progress or transaction. This is the reason why process monitoring is important. SW R&D process should be visible since R&D is very creative and flexible process. So, it is not easy to catch up the progress or change. Previous researches [1, 2] suggest the SW R&D process definition and tailoring with EPF so that standards R&D process can be shareable, customizable and traceable. But it is not enough to make the process to be visible.

Over the last decade business process management (BPM) has become a mature discipline, with a well-established set of principles, methods and tools that combine knowledge from information technology, management sciences and industrial engineering with the purpose of improving business processes [3]. At the first, BPM aims to improve business processes and also is interested in manage, control and

National Research Foundation of Korea (NRF) funded by the Ministry of Science, ICT and Future Planning (NRF-2014M3C4A7030503).

J. A. Kim (✉)
Computer Education Department, Catholic Kwandong University, Gangnung, Korea
e-mail: clara@cku.ac.kr

© Springer International Publishing AG, part of Springer Nature 2019
R. Lee (ed.), *Big Data, Cloud Computing, Data Science & Engineering*, Studies
in Computational Intelligence 786, https://doi.org/10.1007/978-3-319-96803-2_10

support operational processes. Therefore, BPM systems have been emerged. Process models are only useful if they actually help to *improve* processes but models that are sound but at the same time not enough to configure a BPM system do not improve performance. It is the reason we need to automate the process in process execution environment.

In this paper, process execution environment for process defined in EPF is suggested for process visible. From the defined process in EPF, model elements are identified and transformed into the process modeling notation for execution. Section 2 describes the SW process and identified the feature in perspectives of R&D. Section 3 describes the process model for defining the SW R&D Process. This process model is based on the SPEM. Section 4 describes the platform for process management which consists of process definition and tailoring components and process execution component.

2 R&D Application Process Investigation

To redefine the creative and flexible R&D processes, we investigated and analyzed various existing process models to derive the R&D activities that can be considered from the existing process models as well as the R&D related activities that are not provided by the existing processes.

2.1 Agile and Scrum

In the mid-1990s, a new method earlier called as a lightweight method emerged as a lightweight methodology by doing away with the existing heavy and normative methodologies [4, 5]. However, various methods have been called Agile since the release of the Agile Manifesto. Scrum is a typical method. A process based on Scrum, one of the representative practices for Agile, performs an iterative and progressive development. Through Scrum, the product backlog is used to specify the requirements of the development product, and after decomposing it into a sprint backlog, iterative and progressive development is performed by repeating the sprint for one to four weeks. The characteristics of SW research process are as follows:

- Product Backlog: A specification step for requirements and goals of product to be developed.
- Sprint planning meeting: Establishment of development plans for each sprint.
- Sprint: Based on sprint backlog produced through sprint planning meeting, actual research is performed.
- Sprint review: Meetings are held to confirm whether the sprint goal is achieved and what results are obtained.

2.2 Essence Practice

SPEM 2.0, which is a UML-based SW development methodology, was released at object management group (OMG) in 2008 [6]. However, it was difficult to reconstruct various SW development and project management practices, and further difficult to apply this methodology unless it was performed by a SW process engineering expert [7]. Thus, the methodology was not spread throughout the industry. For this reason, a new technique emerged to construct development methodologies by combining latest project practices, which is a type of platform including metal-models and languages released by the Software Engineering Method and Theory (SEMAT). Essence does not have a proprietary process; it is made up of defining processes based on activities that are contained or available in the process. These activity spaces are similar to the components of a typical process, whose order is not separately defined. The characteristics of SW research process are as follows.

- Easy combination of practices
- Easy process custom for project characteristics
- Easy application of new practices

2.3 Unified Process

The Unified Process (UP) is an iterative and incremental development process that can flexibly respond to changes in requirements that would occur during development, as well as changes in the project environment, and can obtain quick feedback from users [8, 9]. The UP has four elements: a worker that defines the actions and responsibilities of an individual or group; an activity that represents the actions performed by the worker; an artifact that the worker creates and controls; and a workflow that describes the order of activities as well as an interaction between the workers. The characteristics of SW research process are as follows:

- Phase-based iterative activities that are suitable for R&D with specific subjects
- Project management workflow, which is a core technology of project control based on project progress and evaluation criteria monitoring activities, as well as managing the iterative plans of UP, which should be essentially analyzed for the integration of new technology development processes.

2.4 Stage-Gate Process Practice

The stage-gate new product development (NPD) process [10, 11] is a system that provides a conceptual and applicable roadmap for the entire process from ideas through new product development to distribution. Each process activity stage is defined as moving to the next stage only after the manager's decision is made through the gate (a type of baseline concept). The characteristics of SW research process are as follows.

- Stage-gate process stage 0: Discovery is the stage of idea development for new products

  - Voice of customer (VoC) activity allows more new features to be discovered
  - Best future scenario/alternative scenario based on ideas are developed and utilized

- When a project is killed at the specific gate, further research is performed by returning to the Discovery Stage.
- New development process of new technologies can be easily integrated into existing NPD process gate.
- For the transition to the next stage, the next action plan is always established.
- The existing NPD process can be scalable according to the characteristics of the project.

2.5 R&D Life Cycle Approach

Numerous process models [12–15] have been proposed for R&D oriented projects, and the traditional spiral model and the unified process have been further reported to support R&D processes. The characteristics of these models are to support conceptualization and iterative product development stages prior to product implementation. The R&D oriented process is largely divided into two types defining the process structure: the first type is the process structure in which the prototype is manufactured and tested through the idea implementation, and the product is developed based on the previous processes, and the second type is the analysis of the possibility through the concrete ideas, gradually developing the final product by using the outcome. The characteristics of SW research process are as follows. The target processes based on R&D include a common process of refining ideas or conducting trade-off analysis and diagnosing the possibility of ideas through the previous steps. Thus, the activities with the following characteristics can be considered.

- Creating ideas: Developing ideas and requirements for developing new products
- Business feasibility analysis: Potential market analysis and application fields for analyzing the merits of ideas
- Product portfolio development: The shape and process of concept on products (prototype + product data)

3   SPEM-Based R&D Process Framework

This research team defined the research & development process framework (RDPF) based on OMG's SPEM 2.0. Accordingly, RDPF is defined based on the RDP meta-model, and the RDP meta-model is defined as a subclass of SPEM 2.0. As a result, the RDP meta-model was defined based on OMG's Model-driven Architecture (MDA), which is defined by Meta Object Facility (MOF) and unified modeling language (UML) meta-model. The RDPF is a tool that authors, tailors, and manages the various processes required to perform R&D tasks on the basis of a formal meta-model. By defining a framework based on a meta-model and further defining a process based on that framework, the consistency, traceability, and reusability can be improved between the activities and the work-products in the intra-process. Furthermore, such framework and process improve the consistency, traceability, and reusability between the development process and management process. The RDPF framework is defined based on the following principles to maximize the effectiveness, productivity, and usability of the framework.

The RDPF meta-model(described in Fig. 1) includes method content and process, and the method content consists of a method class, which is a collection of artifacts and tasks, as well as a method component, which is a collection of method classes. The method class is a reusable unit class centered on work products, and the method component is a structure of method classes, which is typically defined as a unit model. The process is a managerial unit with an operational sequence to achieve milestones. The process component is constructed by sequencing method components by work order, and the delivery process is a combination of process components by work order.

(1) *Method class*: A method class is a set of independent work products as well as activities applied to the products. Typically, a method class defines the artifacts and formats of artifacts that are required to be created for each task of the process, and further defines the guidance for performing tasks or guidance and techniques for producing the artifacts. The method class has an artifact format

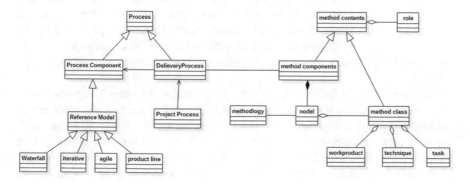

Fig. 1 MetaModel of RDPF

to be created as an attribute, and the operations define the steps to perform tasks or produce an artifact.

(2) *Method component*: A method component is an independent unit that can configure a process according to the software development paradigm and is a basic unit that can constitute a R&D process. A component model includes research model, demand model, analysis model, design model, development model, verification model, and fulfillment model. A method component further refers to a container of model-specific method classes.

(3) *Process component*: A process component is a unit component that constitutes an actual delivery process that reflects temporal concepts and is a composite component that is constructed by assembling already defined method components. A process component is equivalent to a process pattern or a capability pattern described in SPEM. However, the capability pattern of the SPEM has a simple package form, whereas the process component of the RDPF meta-model is encapsulated, having a standardized interface. Thus, it is a process unit that can be commonly reused in defining the delivery process for each of the multiple project. There may be various types of process components. For example, the process component can be an inception component, an elaboration component, a construction component, or a transition component, which are all UP-style process components, or it may further be an agile style process component or a SOA process component.

(4) *Delivery process*: The delivery process is a process that is reflected in the actual project. The delivery process consists of several process components, each of which consists of one or more method components.

4 Platform for RDPF

4.1 Process Definition and Tailoring

The eclipse process framework (EPF) is a tool platform that enables process engineers and managers to implement, deploy, and maintain processes for an organization or individual project. The software development process framework presented by the open source community Eclipse provides SPEM-based process authoring tools to help define the processes, roles, artifacts, formats, and guidance required to perform a project.

Method contents in the EPF are primarily represented by using intermediate artifacts, roles, tasks, and guidance, and supports the definition of guidance such as checklists, examples, or roadmaps. The process primitives of the EPF are overlapped to define a work-breakdown structure (WBS) and are activities that can be interlinked to define a workflow. Such activities further include descriptors that reference method content. Activities can be used to define two basic types of processes: delivery processes and functional patterns supported by the EPF. A delivery process is a fully integrated process template that performs a project of a particular type. A functional pattern is a process to represent and convey process knowledge regarding key areas of interest, such as principles or best practices. The best practices can be effectively reused and applied during the production activities of processes.

4.2 Construction of SPEM-Based RDPF Execution Environment

Although the EPF is a good platform for the definition, deployment, and maintenance of a process, the EPF does not support the concept of running and managing processes.

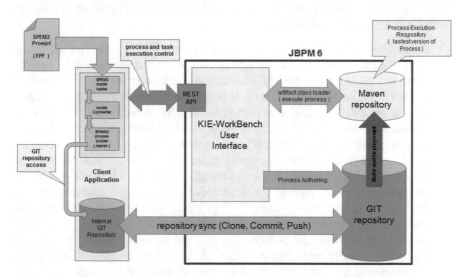

Fig. 2 Architecture of process execution environment

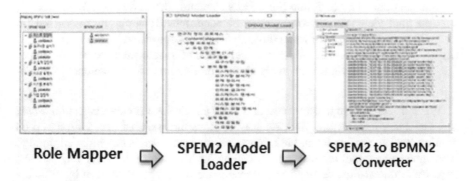

Fig. 3 Model transformation process

Figure 2 shows the architecture of RDPF execution environment. Thus, it is necessary to develop prototyping in conjunction with BPM tools. It is possible to measure and analyze the R&D performance data accumulated while executing the R&D process if the R&D process execution environment platform is constructed that can execute and manage the R&D process by loading the R&D WBS established through the EPF onto the process engine.

The R&D process cannot be executed by the static information of the R&D WBS established through the EPF. Thus, additional technologies are required so that the established R&D WBS can be executed through the process engine, and the R&D performance can be monitored by using the delivery data accumulated during the execution of the R&D process. For this purpose, we identified the components required for the R&D process execution environment platform based on the results of benchmarking the existing process engines (uEngine, jBPM) along with the structure analysis of the WBS created through the EPF, and further selected jBPM as the process engine to be linked with the EPF.

(1) *Conversion of processes constructed based on RDPF into execution model* (Fig. 3): Although the EPF is a platform suitable for process definition, deployment, and maintenance, the EPF must be converted into a model for the process execution engine because the process is not executable. If the EPF-based process model and the model that is executed by the currently available process execution environment are corresponding or compatible, then the model conversion is possible. However, that is not the case in the real world. For example, in the typical process execution environment, there is no concept for management of iteration, role, and artifact. Through this study, it became possible to add process iteration concept, role concept of process engine registration user, document unit and management concept between RD unit artifacts into the execution environment. Furthermore, we have extended the process to select the standard activities and tasks to be performed for each iteration. We defined the correspondence between the process model defined by the SPEM standard and the model defined by the BPM engine and designed the model converter. The following figure defines the procedure to convert the SPEM-based process model defined by EPF into the BPMN standard. Based on the refined process, the final view of the process model loaded onto BPM can be visualized (in Figs. 4 and 5) and confirmed through the model converter.

(2) *Verify the transformed process*: Visualization of transformed process (in Fig. 6) helps the process modeler to check the correctness and completeness for execution.

When a task is assigned to a researcher by executing a process based on a user tray and agent architecture, the user tray displays the assigned task with a pop-up window, and an artifact format relating to the assigned task is downloaded through the tray for work, and the task results can be further uploaded. Moreover, we have improved the real-time demand and convenience of process execution by allowing the researcher to decide whether all tasks allocated to the researcher would proceed and whether they have been completed by a tray approach. The events on the client

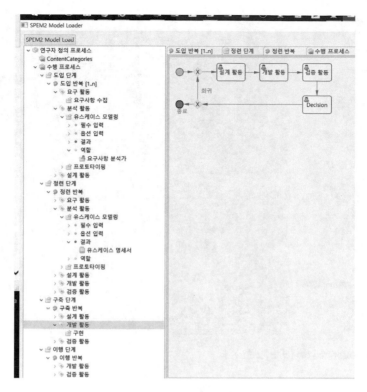

Fig. 4 Transformed model in process modeling notation

for task execution by each researcher are aggregated to the server through git, and the implementation includes that the administrator can monitor the entire research progress via the server.

5 Conclusion

The present study established the RDPF to provide a reference model for defining the project-specific process that should be established for software development. For this purpose, we analyzed existing R&D processes and practices, and further analyzed SPEM model and EPF for standardization. Based on the standard model, this study proposed a meta-model of the SW RDPF, and further defined the components of the process framework based on the meta-model.

Despite various existing software process models, a case study was conducted focusing on the development activities required in the R&D process from the perspectives of the comprehensive project management, rather than the engineering process of simple software development. Through the analysis, we summarized what

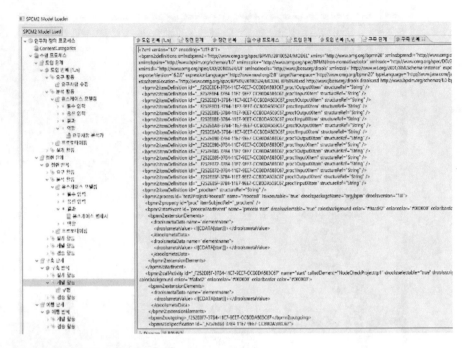

Fig. 5 Transformed model in XML

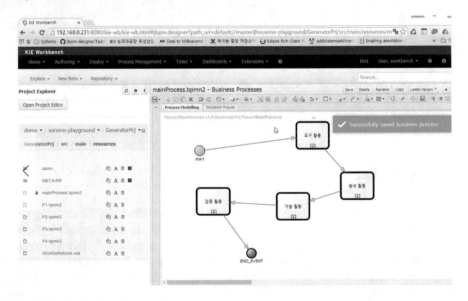

Fig. 6 Example of model execution

activities can support R&D characteristics upheld by existing processes and practices, and what research descriptors (RDs) can be defined from the activities. The SPEM model and the EPF were analyzed as an environment for construction of a standard process model as well as establishing and tailoring the process. SPEM is a meta-model developed by the OMG to define the software development process and related items, such as terms, concepts, and relations, which defines a set of essential concepts for modeling, exchange, enactment, documentation, presenting and management of a development method or process. The SPEM model was analyzed and the RDP meta-model was re-defined based on OMG's MDA, which is defined by MOF, UML, and meta-model. This makes it possible to guarantee scalability and reusability. The EPF is a software development process framework provided by the open source community Eclipse is a tool platform for process engineers and managers to implement, deploy, and maintain processes for organizations or individual projects. The EPF provides a SPEM-based process authoring tool, which can easily define the processes, roles, artifacts, formats, and guides required to perform a project. The RDPF defines through the EPF a platform that authorizes, tailors, and manages the various processes required to perform R&D tasks on the basis of a formal meta-model. We verified that it was possible to create a delivery process customized for each project by deploying the elements defined in RDPF in EPF to confirm the feasibility of the RDPF. The RDPF provides a process monitoring environment through a process execution environment, and further provides a basis for efficient process management.

Acknowledgements This research was supported by Next-Generation Information Computing Development Program through the National Research Foundation of Korea (NRF) funded by the Ministry of Science, ICT & Future Planning (NRF-2014M3C4A7030503).

References

1. Choi, S.Y., Choi, J.Y., Kim, J.A., Choi, J.Y., Cho, Y.H., Hong, J.E.: Process tailoring practice with EPF. In: The 10th Asia Pacific International Conference on Information Science and Technology, APIC-IST 2015, pp. 115–120 (2015)
2. Choi, S.Y., Choi, J.Y., Kim, J.A., Choi, J.Y., Cho, Y.H., Hong, J.E.: Software R&D process tailoring practice with EPF. Int. J. Appl. Eng. Res. **10**(5), 3979–3982 (2015)
3. van der Aalst, W.M.P., La Rosa, M., Santoro, F.M.: Business process management. Bus. Inf. Syst. Eng. **58**(1): 1–6(2016)
4. Schwaber, K., Beedle, M.: Agile software development with SCRUM, Prentice Hall (2002)
5. Olsson, H.H., Bosch, J., Alahyari, H.: Towards R&D as innovation experiment systems: a framework for moving beyond agile software development. In: The Proceedings of the IASTED (2013)
6. OMG: Software & Systems Process Engineering Meta-Model Specification Version 2.0 (2008)
7. Ng, P.-W., Huang, S.: Essence: a framework to help bridge the gap between software engineering education and industry needs. In: IEEE 26th Co Conference on Software Engineering Education and Training (2013)
8. Ivar, J., Grady, B., James, R.: The unified process. IEEE Soft. **16**(3), 96–102 (1999)
9. Kruchten, P.: The Rational Unified Process: An Introduction. Addison-Wesley (2003)

10. Cooper, R.G., Edgett, S.J.: Optimizing the stage-gate process: what best-practice companies do. Res.Technol. Manag. **45** (2002)
11. Grönlund, J., Sjödin, D.R., Frishammar, J.: Open innovation and the stage-gate process: a revised model for new product development. Calif. Manag. Rev. **52**(3) (2010)
12. Bandarian, R.: From idea to market in ripi: an agile frame for NTD process. J. Technol. Manag. Innov. **2**(1) (2007)
13. Chaochotechuang P., Daneshgar F., Sindakis, S.: Innovation strategies of new product development (NPD): case of thai small and medium-sized enterprises (SMEs), In: Sindakis, S., Walter, C. (eds.) The Entrepreneurial Rise in Southeast Asia. Palgrave Studies in Democracy, Innovation, and Entrepreneurship for Growth. Palgrave Macmillan, New York (2015)
14. Jain, R.K., Triandis, H.C., Weick, C.W.: Managing research, development, and innovation. Wiley (2010)
15. Astebro, T.: Key success factors for technological entrepreneurs' R&D projects. IEEE Trans. Eng. Manag. **51**(3) (2004)

A Study on the Factors Affecting Intention to Introduce Big Data from Smart Factory Perspective

Won-Jung Jang, Soo-Sang Kim, Sung-Won Jung and Gwang-Yong Gim

Abstract In order to strengthen the competitiveness of the manufacturing industry, major developed countries are actively promoting smart factory construction. Smart Factory is a factory that can install sensors on elements of physical factories, convert physical signals into digital signals, connect machines, people, supply chain partners, etc., and smartly manage them on their own based on collected big data. The core of Smart Factories is at its power of data analysis, and the introduction of Big Data, which is the base for this, is very important. In particular, the percentage of SMEs in Korea is very high (99.82%), and in order to strengthen the competitiveness of manufacturing industry, the policy for introduction of SMEs 'smart factory is continuously implemented, but, the level of smart factory introduction stays in its basic level. The core to smart factories is that, although the introduction of big data is very important in the power of data analysis, the research about examining the cause about the influence on introduction is very insufficient. In this paper, it is meaningful that we deducted and suggested the factors affecting the intention of big data introduction.

Keywords Big data · Smart factory · TOE framework
Smart factory framework · Manufacturing

W.-J. Jang
Department of Intellectual Property for Startups, Catholic Kwandong University, Gangneung, South Korea
e-mail: wjjang@cku.ac.kr

S.-S. Kim · S.-W. Jung
Department of IT Policy and Management, Soongsil University, Seoul, South Korea
e-mail: sskim@comtec.co.kr

S.-W. Jung
e-mail: jdate@naver.com

G.-Y. Gim (✉)
Department of Business Administration, Soongsil University, Seoul, South Korea
e-mail: gygim@ssu.ac.kr

© Springer International Publishing AG, part of Springer Nature 2019   129
R. Lee (ed.), *Big Data, Cloud Computing, Data Science & Engineering*, Studies
in Computational Intelligence 786, https://doi.org/10.1007/978-3-319-96803-2_11

1 Introduction

Smart Factory is a factory that can itself operate with intelligence which connects all the things, people and cooperation partners in the production site in real time and raises the productivity by predicting the operation, inspection, and maintenance of the machine through the analysis of collected big data, and optimizes the production and operation of the plant. Major developed countries around the world are making efforts to strengthen their manufacturing competitiveness, such as policies and programs [1]. Considering that manufacturing accounts for 28% in China, 12% in the U.S., 23% IN Germany, and 19% in Japan, Korea is very high with 30% in manufacturing [2]. Our country divides companies into large company, middle-sized company, small company, and small business, and as of 2014, and the total number of enterprises in Korea is 397,171, with 3,957,394 employees [2]. The number of large and middle standing companies is 701 with 0.18%, and that of small and medium enterprises (SMEs) is 396,470 with 99.82%, and 3.22 million persons with 81.4% are working in small and medium manufacturing companies, playing a pivotal role in manufacturing industry [2]. As of December 2016, Korea has 2,800 smart factories and aims at supplying and advancement of 20,000 of it by 2022 [3]. The level of smartization can be divided into four categories: basic, middle 1, middle 2, and advanced. Of the 2,800 smart factories, 79.1% are in basic category [3]. According to the Report of Smart Factory Promotion Status Survey by Korea Industry 4.0 Association and KMAC in 2016, the solution to be first introduced to production site for implementation of the Smart Factory emphasizes the introduction of Big Data with Big Data (21.9%), industrial robot (15.7%), ERP (12.3%), 3D printing (11.5%), MES (10.4%), sensor (10.1%) and cloud (6.4%). The characteristics of big data are volume, variety, velocity, and veracity of data, and securing data reliability is even more important [4]. The core of Smart Factories is the power of data analysis and Big Data is the starting point for this. The purpose of this paper is to deduct and suggest the factors affecting the introduction of big data, which is the core of smart factory introduction and advancement.

2 Paper Preparation

2.1 Technical Factors that Affect Smart Factory Construction

A smart factory means a factory that itself takes care and operates of a factory smartly. The Smart Factory defined by the Public and Private Smart Factory Promotion Group is a factory that improves productivity, quality, and customer satisfaction by applying information and communication technology to manufacturing processes such as planning, design, production, process, distribution and supply chain management. In terms of technology development for smart factory construction, smart factory design/operation/optimization, machine learning, manufacturing intelligence based

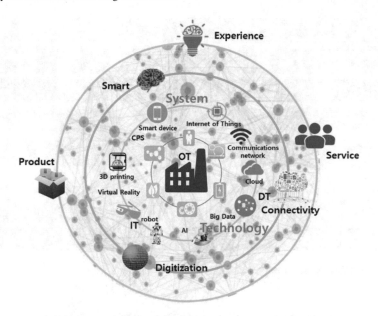

Fig. 1 Factors affecting smart factory construction

on artificial intelligence, Cyber Physical System (CPS), Industrial Internet of Services (IIoS), Industrial Internet of Things (IIoT), Industrial Data Analytics, Smart Sensors, Intelligent Robots/Automation Facilities, Interoperable Platforms, Cloud Computing, Manufacturing Intelligence, and other core technologies are being researched and developed [5]. Smart manufacturing is a strategy to optimize all production processes, and the core execution strategy of Smart Factory is that Smart factory CPS manages, refines, and analyzes big data collected at the manufacturing site, establishes a cyber-model that matches the situation in the field through real-time synchronization, achieving operational optimization by itself [5]. The Smart Factory is connected horizontally, combined, and operated autonomously with facility, process management, manpower, and supply chain based on Operation Technology (OT), Information Technology (IT), Data Technology (DT) [2]. Because the Smart Factory is connected in seconds, it provides customized value to individuals and companies through product and service experience, and can improves productivity and efficiency and enables optimal operation. Figure 1 is a factor affecting the establishment of smart factories [2].

2.2 Framework for Building a Smart Factory

In the era of the 4th Industrial Revolution, it is very important to provide the customized value of the enterprise or individual that the customer wants to maintain

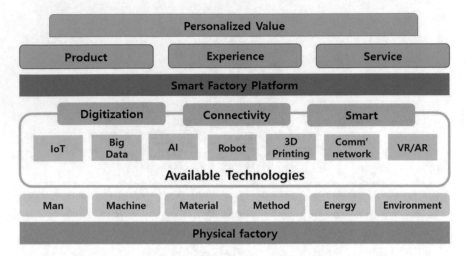

Fig. 2 Smart factory framework

the competitiveness of the enterprise. In order to continuously provide customized value to the enterprise through the products, services, and customers' experiences that the company produces, it is necessary to lower the manufacturing cost to the desired level, secure the quality and productivity, and operate the factory flexibly and smartly by itself. Figure 2 is smart factory framework [2].

With 4M2E (Man, Machine, Material, Method, Energy, Environment), the element of the physical factory at the manufacturing site, it responds to the direction required by the time according to technical development and environmental change [2, 6]. It can digitize physical signals to enable communication by attaching sensors to old machinery or equipment, connect machines, parts, factories, manufacturing processes, supply chain partners, and people in the factory to each other, and the smart factory platform smartly operate based on the collected big data [2]. IoT, big data, and artificial intelligence are very important technical elements for smart factory construction.

2.3 Related Research of Big Data

The introduction of the Big Data System requires a systematic roadmap, and it is said that as a factor to introduce Big Data, it starts with a rapidly changing technological perspective, with the need to create new data types and new information processing [7]. The introduction of Big Data for building a Smart Factory at the production site is very important. The big data introduction process has three steps, and its consideration by step is as in Fig. 3 [2].

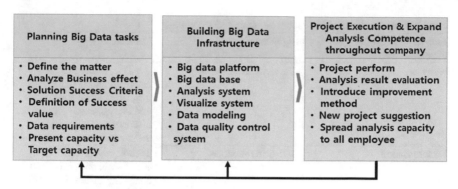

Fig. 3 Big data introduction process

The purpose of this study is to help identify the essential factors and influences related to the introduction of big data, which is very important for smart factory construction. However, there has been a lot of interest in introducing smart factories in recent years, and the introduction of big data has to be preceded for the introduction of smart factories, but related researches are not enough. Looking into the study on the introduction of Big Data, management support in introducing new systems greatly affects in applying in-house business processes for employees introducing of technology and strengthening of competitiveness [4]. And, it affects the system introduction of competitors in the same industry, and it is because the introduction of new technologies by competitors may give fear to customers that they will fall behind competition. The element to introduce big data system starts with company's desire to expand its business [7]. Companies are interested in introducing and using big data systems to be competitive. Because the risk increases if companies fail to comply with data management regulations, they have been warned of the need for data management through institutional support for establishment of business rules and ethics [8]. And the necessity of data control due to the increase in issues related to the storage and analysis of big data is suggested due to the increase of data usage [9]. Cost efficiency is an important factor for new investment in companies, and ROI, which means investment return, is considered as an essential consideration [10].

2.4 A Study on the Introduction Model of Information System

It is suggested that the factors of acceptance of Big Data System technology, which is an innovative technology, are suitable for describing the cost and business process changes in the difficult technical elements such as big data analysis technology, open source technology and introduction through the TOE model. In order to study the acceptance characteristics of the domestic big data technology, this study integrated the Diffusion of Innovation (DOI) and the Task Technology Fit (TTF) based on the Technology Acceptance Model (TAM) [11]. This study defined Big Data Technol-

Table 1 Characteristics of big data as an innovative technology

Innovative technical characteristics	Big data relevance	Characteristics
Newness	Creativity of technology	Big Data Technology is a newly emerging innovation through creative approaches and these technologies are not disconnected but connected and fused to create new value
Differentiation	Data differentiation	Techniques for collecting, combining, processing and analyzing various types of regular, irregular, and semi-regular data
Externality	Fusion of data	Opening government-led public data and creating external effects through the purchase of necessary data
Economic value	Create economic effect through new business	Optimize existing business with big data and create new economic profit through new business model

Source Reorganization based on the contents of Kim (2014)

ogy as an innovative technology because Big Data has characteristics of innovative technology. The characteristics of innovation in Big Data are described as Newness, Differentiation, Externality, and Economic value, and its details are as in Table 1 [11].

The innovative technological factor of Big Data is the background of the use of the innovation diffusion theory variable in this study.

2.4.1 Technology/Organization/Environment Framework

The TOE Framework has been proposed to arrange the adoption of innovation information technology and understand the factors that affect the introduction of new information technology [12]. Organizational IT related variables are described from three viewpoints: Technology Context, Organization Context, and Environment Context. The technical factor (T) is the available technological factor that exists in inside/outside of the company, and the organizational factor (O) is the factor measured with the size of company, the degree of centralization, the complexity of management organization, quality of human resources, and spared resources, and environmental factors (E) are factors affecting the outside of the organization, including industry, competitors, governments, and suppliers of resources [13]. The TOE framework is mainly used to the studies that describe the technology acceptance, homogenization, and diffusion phenomenon in the enterprise aspect [14]. The TOE framework was

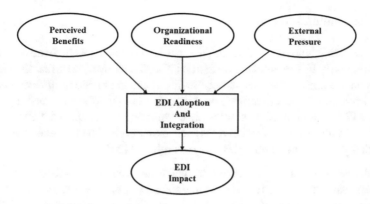

Fig. 4 EDI acceptance model by Iacovou et al

first applied in the information security field, the big data field, the e-business field, the open system field, and the EDI-related empirical studies [15–19]. The benefits perceived by low use in EDI in SME are limited through the study of external pressure as for the element of preparedness of organization, environmental factors for technical benefit, organizational element perceived as technical element in Fig. 4 of a study on acceptance and diffusion of EDI, and external pressures from outside, such as insufficient of preparing of organization with low market standing in technical introduction work as major factor [13, 17].

The Technology-Organization-Environment (TOE) framework has proven its effectiveness in research models of various fields and provides a continuous theoretical basis. In particular, researches have been conducted in various information communication technologies field such as cloud computing service, big data, and e-business [20]. In a study about determinants of Big Data introduction, cost, security concern, relative advantage, compatibility, complexity, top management support, competitor pressure, and regulation support were adopted as the technical factors of the TOE framework [18]. From the viewpoint of resource base, security, data management and IT infrastructure were adopted as technology factors of TOE in the study about the intention of using Big Data. In this study, the security factors were adopted as the technical factors of the TOE, and the task technology suitability factors were adopted as the environmental factors. In a study about the factors affecting the intention to introduce big data, reliability, appropriateness, and ease were adopted as the technical factors of TOE framework, and CEO, organizational attitude, company characteristics, and organization informatization were adopted as organizational factors, and inside/outside environment of the industry, public environment were applied, and as a result of empirical study, it is suggested that the technical factors affect the intention to introduce big data [15, 20, 21]. Thus, it was confirmed that the technology, organization, and environmental factors of the TOE framework have a significant impact on the organization's intention to accept big data and the diffusion of new technology.

2.4.2 Innovation Diffusion Theory

It is assumed that innovation diffusion theory can be used to test innovative technologies, particularly ICT technology diffusion [22]. The innovation diffusion theory has focused on identifying the reasons for the difference in the rate of diffusion of innovation defined as "ideas, practices, or things that are perceived as new" by society members [23]. And empirical studies have examined the effects of perceived characteristics on innovation adoption rates most extensively. The five main variables of innovation diffusion theory are described as follows [24].

(1) Relative Advantage: The degree to which innovation is considered "better for users than for existing ideas (products and services)". The examples of ideas are the existing economic benefits, social reputation, convenience and satisfaction. The relative advantage is not absolute, but depends on the perception and needs of the user group.
(2) Compatibility: Innovation is "the degree to which existing values, experiences and needs are perceived as continuity from user groups". If new innovation factors are inadequate to existing values, they will not be accepted rapidly.
(3) Complexity: New ideas can be transformed by active users, and complexity is "the degree to which innovation is perceived as difficult to understand and use". "The concept of re-invention that continues to evolve is the key to expanding innovation".
(4) Trialability: "The degree to which innovation can be experienced or tested beforehand." It is a plan in which the risk of new products or services is to be minimized before adoption.
(5) Observability: "If the results of innovation adoption can be identified, users can more easily select innovation ideas". It can reduce uncertainty and facilitates the introduction by sharing and identifying information with surrounding stakeholders.

Among the five key variables of innovation diffusion theory, relative advantages, complexity and compatibility are also used as technical variables in the technology, organization, and environment frameworks. The probabilities and observability are excluded from this study because they can be done after the introduction of Big Data.

3 Paper Preparation

3.1 Research Model

In this study, introduction of big data for building smart factories is very important, but it is not done well, so the TOE model to test the factors of big data introduction is suggested. The research model is established as in Fig. 5. As a result of previous studies, it was found that technical factors (cost, security, relative advantage, com-

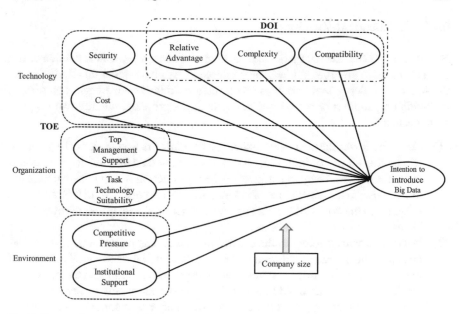

Fig. 5 Proposed research model

plexity, compatibility), organizational factors (support of management, suitability of technical skill), environmental factors (pressure of competitors, institutional support) company size, and type of manufacturing industry were used as control variables in the relationship between all factors and the intention of company to introduce big data is set as dependent variable.

The reason for adopting the TOE model in this study is that the big data technology is expensive to introduce open source technology, various technology factors, analysis technology, data leakage risk, and introduction and the technical factor of TOE model is suitable as there can be change in work process. Also, introduction of big data is essential due to necessity of man costs, and the organizational factors of the TOE model are suitable because big data can be used effectively to derive effective results in certain business areas with clear goals. The environmental factors of the TOE model are appropriate because it is possible to decide to introduce big data due to the fear that it can fall behind competition upon competitor's introduction of big data as well as institutional support for building a smart factory to strengthen the manufacturing competitiveness of the government. The reason for adopting the DOI model, which is used in research on the acceptance of innovation technology, is that it needs to be included in the research model composition because Big Data contains the characteristics of innovation technology. Since the DOI variable suitable for the introduction of big data can be included in the TOE technology factor, the research model is composed of the TOE model. In this study, the influence of all factors was compared with the firm size. The reason for the comparison is that it is judged there will be differences in introduction factors depending on the size of the company.

3.1.1 Independent Variables

Based on the TOE and DOI theory and the literature related to the introduction of Big Data reviewed in previous studies, the proposed research model of this study was set up and the 9 independent variables (cost, security, relative advantage, complexity, compatibility, support of management, competitor's pressure, institutional support) were deducted.

(1) Cost: The lower the cost of introducing a new information system, the sooner it will be introduced or built [25]. As the cost of big data systems introduction is high, it will have a negative impact on the introduction if cost is high compared to value due to the acceptance of information technology [18]. In this study, it is judged that the cost will have a negative impact on the intention of introducing big data.

(2) Security: Security refers to the control of data access inside/outside the enterprise and the degree of external leakage of data. It is argued that the degree of internal/external security systems of an enterprise is an important factor in using big data [26]. It is assumed that security concerns about new IT investment are important factors [10]. And there are also a number of security concerns about the leakage of internal/external data in the introduction of Big Data. This study is going to analyze data security concerns.

(3) Relative Advantage: It is said that relative advantages are perceived as being lower than what innovation does not [23]. When it is believed that the innovation of the companies will increase efficiency and help obtaining economic benefits, they will introduce innovation [27]. Relative advantage was considered as a positive and important variable related to the introduction of information technology innovation [19]. Relative advantage is said to be an important variable to the spread of big data introduction. The relative advantage in this study is judged to have a positive effect on the intention of big data introducing [18].

(4) Complexity: Complexity means that innovation is relatively difficult to use [23]. It is said that it is a factor that determines the acceptance of the information technology depending on the user's perceived difficulty and complexity of the new information technology [28]. As information technology complexity increases the uncertainty of successful implementation, it increases the risk of decision making of introduction [29]. Complexity gave a positive impact on the intention of introducing big data [18]. In this study, if the business process and usage learning by using big data is more convenient and easier than existing information, it will affect intention of introduction, so it is judged to be an important variable of the intention of big data introduction.

(5) Compatibility: Compatibility includes not only mechanical compatibility, but also compatibility with existing values, past experiences, and organizational needs, both inside and outside the organization. In the analysis of prior studies on compatibility, only relative advantages, complexity, and compatibility are consistently explained as the major influencing factors on innovation acceptance [30]. When adopting information technology, the acceptance organization con-

siders whether compatibility meets the organizational needs and procedures of new information technology. In this study, compatibility considers conformity to the existing value and present necessity by using Big Data as an important variable affecting the intention to introduce Big Data.

(6) Top Management Support: CEO support is the degree of support provided by senior management in introducing technological innovation in business [31]. The support of CEO has a beneficial relationship to the choice of the company for innovation acceptance, and it has also been proposed in the study of IT innovation acceptance based on the TOE framework [31]. It is easy to introduce innovative technologies as long as there is a firmer CEO support for innovation systems [32]. The support of the management is essential because many people in the IT and non-IT departments are required to be input, educated in the introduction and utilization of big data, and there will be difficulties due to conflicts in opinions between departments [18]. In this study, the support of CEO top management is considered to be an important factor affecting the intention of introducing big data based on previous studies.

(7) Task Technology Suitability: Task technology suitability refers to the degree of suitability with departments and personal affairs in introducing Big Data. And the extent to which IT properly supports the functions that individuals need to perform their tasks [33]. IT can be appropriately applied to new IT utilization, depending on how well it is suitable to the requirements or requirements of their work existing individuals perform [33]. Companies that are planning to introduce big data should introduce it with a clear purpose in big data task planning. This study is going to analyze task technology suitability as for the effectiveness of individual work through utilization of Big Data in terms of TOE organizational factors.

(8) Competitive Pressure: The environment is the place where a company conducts its business, and is influenced by its competitors, the nature of its industry, the smooth use of resources supplied from others, and its interaction with government [34]. Competition pressure has been recognized as an important driving force for technology diffusion in literature related to innovation diffusion, and is said to be the pressure from competitors in the industry [35–37]. In competitive organizations the company can be relatively competitive if they introduce big data first and use it for business, leading the industry and using it as a marketing factor [18]. In this study, competitive pressures between firms are considered to be important variables influencing intention to introduce new innovative technologies.

(9) Institutional Support: Institutional support is the support of government authorities to encourage companies to be assimilated with IT innovation [38]. The introduction of new information technologies will have a positive impact on the introduction of Big Data if the regulations are well established [18]. With the increase of big data, laws and institutional regulations on data transactions are becoming important. In this study, institutional support of policy authorities is considered as an important variable that can influence the intention of big data introduction.

3.1.2 Control Variables

(1) Company size: The difference in company size affects the intention of big data introduction [21]. In a previous study of company size variables, it is assumed that "the introduction of the system requires greater technology and capital, and could be introduced more easily by companies with large size, and the size of the company is an important variable in the introduction of ICT" [39]. Currently in our country the company size is divided into large company, middle standing company, small company, and small business, and this study divides it into large company, middle standing company (more than 300 persons), middle sized company (50–299 persons), small company (<50 persons) [2].

3.1.3 Dependent Variables

(1) Intention to introduce Big Data: The intention to introduce means the degree of activity that is being carried out for introduction and utilization of big data. Companies that are considering introducing Big Data should clearly understand the strategic value that can be gained through the introduction of Big Data, and it is desirable to review and introduce the aspects of the internal environment of the industry [21]. This study is going to select and analyze the intention of introduction of big data as a measure of success in introducing and using it [40, 41].

3.2 Hypothesis Setting

Hypothesis <H1>	Cost has a negative (−) impact on intention to introduce big data.
Hypothesis <H2>	Security has negative (−) impact on the intention to introduce big data.
Hypothesis <H3>	Relative advantage has a positive (+) impact on intention to introduce big data.
Hypothesis <H4>	Complexity will have a positive (+) impact on intention to introduce big data.
Hypothesis <H5>	Compatibility will have a positive (+) impact on intention to introduce big data.
Hypothesis <H6>	Management support will have a positive (+) impact on the intention to introduce big data.
Hypothesis <H7>	The task technology suitability will have a positive (+) impact on the intention of introducing big data.
Hypothesis <H8>	The competitive pressure of will have a positive (+) impact on the intention of introducing big data.

Hypothesis <H9>   Institutional support will have a positive (+) impact on the intention to introduce big data.

Hypothesis <H10>   The impact of technical variables, organizational factors, and environmental factors on the intention of introducing big data will be adjusted according to the firm size.

3.3 Operational Definition of Variable

The Operational definitions in this study are shown in Table 2.

4 Hypothesis Testing and Analysis Results

4.1 Characteristics of Study Sample

This study conducted an empirical analysis targeting the members of Korea Industry 4.0 Association in order to investigate the factors that affect the intention of introducing big data, as there is a lot of interest in building a smart factory, but actual introduction of Big Data is not done properly. In order to increase the reliability and accuracy of the questionnaire, this study conducted a questionnaire survey targeting the association members who have a high awareness of Smart Factory and Big Data. This study collected 102 valid questionnaires through e-mail, visit, and online and conducted statistical analysis based on them. The respondents who introduced Big Data were 30 (29%) and those who did not introduce Big Data were 72 (71%), a relatively large number of respondents. In terms of business types, information technology was the most with 49 (48%), and 19 in manufacturing/distribution, 49 in information technology, 13 in public/government, 6 in education/consulting, 11 in banking/insurance, and 4 others. The size of the enterprise was the most with 48 in large/middle standing, 30 medium-sized companies and 24 small companies. The size of yearly sales was the most in 38 persons in below 30 billion (38.7%), with 19 persons in over 1 trillion, 3 persons in 500 billion–1 trillion, 22 persons in 100–500 billion, 20 persons 30–100 billion. The position was the most in 40 executives (39%), followed by 26 managers (25%), 22 section chief/deputy manager (22%), 9 employees (9%) and 5 assistant manager (5%). The age was 1 person in over 60 years old, 34 persons in 50–59 years old, in which 40–49 years old accounts for 48% of the total. The number of years in service was the most with 37 persons in over 20 years), followed by 35 persons in 10–20 years, 17 persons in less than 5 years. There were 33 respondents of having (32%) and 69 of not-having (68%) in the big data organization. There were 24 respondents of having (23%) and 78 of not-having (77%) big data policy regulations.

Table 2 Operational definition of variables

Variable		Operational definition
Technical factors of the TOE	Cost (CT)	The cost consideration factor upon big data introduction, in the part of big data implementation, operation, installation, education and management
	Security (SC)	The security is defined as the degree of control and outside leakage of data access
Common factors between the technical factors and DOI of the TOE	Relative advantage (RT)	The degree to which innovation is recognized as better than the existing technology by introducing big data
	Complexity (CX)	The degree of perception that it is relatively difficult to understand or use by users in comparison with existing technology in accommodating big data
	Compatibility (CB)	The degree to which innovation is compatible with existing values and present needs by using Big Data
Organizational factors of the TOE	Top management support (TM)	The level of awareness and support of CEO on the introduction and utilization of Big Data
	Task technology suitability (TS)	The degree to which service and function related to big data were perceived as suitable for carrying out tasks
Environmental factors of TOE	Competitive pressure (CP)	The degree of change in competition between companies by introducing big data in business activities
	Institutional support (IS)	The degree of support for information protection regulations and government legislation that encourage the introduction of big data
Intention to introduce big data (BI)		The degree of activity that the company is undertaking to introduce and utilize big data

4.2 Reliability and Validity Analysis of Measurement Tools

In this study, PLS analysis technique was used to verify the hypothesis of the research model. PLS is advantageous for exploratory research without a normal distribution constraint on sample size, variables, and residuals based on Principle Component Analysis [42]. PLS also has the advantage of reducing the prediction error by using the least squares method instead of the maximum likelihood method. PLS uses Composite Reliability and Average Variance Extracted (AVE) values similar to Cronbach's Alpha values for the reliability analysis of measurement items [43]. The absolute correlation between the latent variable and its measurement variable should be 0.7 or higher to evaluate the internal consistency [44]. If the mean variance extracted value is 0.5 or more, the reliability of the measured variables is considered to be secured. As a criterion for the factor analysis of this study, the factor load value of the exploratory factor analysis was applied to the case of 0.5 or more which is universally significant in the field of social science. Table 3 is the exploratory factor analysis, and IS4, IS5, CX1, CB4, BI4, BI5, CP1, and CP5 variables are removed.

The results of the reliability analysis are shown in Table 4.

Intensive feasibility should allow measurement items to be expressed in their single dimensionality. It can be viewed that there is discriminant validity in case of suggesting the use of average variance extraction as a criterion of convergent validity, and when it exceeds or over 0.5 of the squared value of the correlation coefficient between constructive concepts [45]. This model meets the criteria for reliability and discrimination and is the same as Table 5.

4.3 Hypothesis Testing

The interpretation of analysis result of the structural equation model of PLS is measured by determination coefficient (R^2) of the size of the path coefficient, the sign, the statistical significance, and the final dependent variable described as independent variables. The hypotheses of this study was done by using the path coefficient of the PLS structure model and by verifying the significance of the paths between the variables by using the measurement model for the factors that examined the reliability and validity of the measurement items. The t-value was calculated by generating 500 repeated extraction sub sampling through bootstrapping, which is a method of estimating a measurement with the same distribution by reconstruction extraction from the sample data. In this study, t-value was used for the significance level evaluation. Over 1.96 (P-value < 0.05), which is significant in the social science field, is applied as reference value. The results of each hypothesis were examined by using the path coefficient of the PLS structural equation model with technical, organizational, and environmental characteristics. Path coefficient in case of costs, security, relative advantage, complexity, and compatibility for technical characteristics, hypothesis 1, 2, 4 were rejected with value for costs 0.127(3.247), security −0.127(1.702), relative

Table 3 Exploratory factor analysis

Variables	Component										Cronbach α
	1	2	3	4	5	6	7	8	9	10	
TM3	0.866										0.915
TM2	0.832										
TM4	0.827										
TM1	0.747										
TM5	0.720										
SC2		0.923									0.944
SC4		0.917									
SC5		0.888									
SC3		0.887									
SC1		0.794									
TS2			0.831								0.899
TS3			0.818								
TS4			0.789								
TS1			0.696								
TS5			0.663								

(continued)

Table 3 (continued)

Variables	Component										Cronbach α
	1	2	3	4	5	6	7	8	9	10	
CT5				0.829							0.838
CT4				0.814							
CT2				0.770							
CT3				0.709							
CT1				0.636							
IS2					0.858						0.951
IS3					0.856						
IS1					0.838						
RT3						0.777					0.844
RT2						0.752					
RT4						0.656					
RT1						0.565					
CX4							0.878				0.814
CX3							0.773				

(continued)

Table 3 (continued)

Variables	Component										Cronbach α
	1	2	3	4	5	6	7	8	9	10	
CX5							0.756				
CX2							0.603				
CB3								0.824			0.888
CB2								0.796			
CB1								0.618			
BI3									0.734		0.895
BI2									0.724		
BI1									0.618		
CP3										0.680	0.807
CP4										0.597	
CP2										0.510	

Table 4 Results of the reliability analysis

Variables	AVE	C.R.	R^2	Cronbach α	Communality	Redundancy
BI	0.826	0.9348	0.607	0.895	0.826	0.155
CB	0.818	0.931		0.888	0.818	
CP	0.727	0.888		0.812	0.727	
CT	0.564	0.864		0.855	0.564	
CX	0.623	0.868		0.815	0.623	
IS	0.912	0.969		0.952	0.912	
RT	0.687	0.897		0.847	0.687	
SC	0.727	0.929		0.943	0.727	
TM	0.749	0.937		0.916	0.749	
TS	0.715	0.926		0.900	0.715	

advantage 0.112(2.233), complexity −0.146(4.419), and compatibility 0.200(4.615), and hypothesis 3, 5, and hypotheses 3, 5 were adopted. For the path coefficient value for the support of the management, task technology suitability that fall on organizational characteristics, the hypotheses 6, 7 were adopted with 0.104(2.781) for the management support and 0.137(3.055) for the task technology suitability. Path coefficient value for pressure of competitor, institutional support corresponding to environmental characteristics, hypothesis 8, 9 were adopted with value of 0.320(7.777) for pressure of competitor, and 0.146(3.600) for institutional support. Also, as independent variable and the value (R^2) of determination coefficient for the model about Big Data introduction intention in this structural model is 0.607, whose explanation power can be viewed to be great.

The bootstrap resampling technique means random sampling to determine the accuracy of the prediction statistics and generating a statistical distribution from each sampling [46]. The bootstrap technique is mainly used to evaluate significance of path coefficient in PLS path model [47]. The summary of analysis result in PLS path in this study is as Table 6.

Total model fit of PLS path model in this study was confirmed to be very good with 0.668. The results of model fit analysis is as Table 7.

As a result of the control effect analysis, it was confirmed that there are differences between the two groups in the competitive pressure, cost, institutional support, security, and management support. The results of the analysis of the adjustment effect in company size between large/middle standing companies (48 persons) and small companies (24 persons) are shown in Table 8.

4.4   Results of Analysis

First, hypothesis test results between technical characteristics and intention to introduce big data are analyzed as follows. The effect of the cost on the intention of

Table 5 Correlation coefficient between variables and square root value of AVE (discriminant validity)

BI	CB	CP	CT	CX	IS	RT	SC	TM	TS
0.826									
0.572	0.818								
0.689	0.566	0.727							
0.235	0.206	0.128	0.565						
0.251	0.461	0.316	0.202	0.623					
0.534	0.350	0.579	0.263	0.227	0.912				
0.559	0.535	0.547	0.297	0.389	0.360	0.687			
−0.106	−0.050	−0.084	0.364	−0.054	0.067	−0.010	0.727		
0.527	0.412	0.577	0.101	0.287	0.549	0.438	−0.050	0.749	
0.509	0.530	0.512	0.146	0.476	0.319	0.625	0.005	0.363	0.715

Table 6 Statistical analysis results of hypothesis verification

Hypothesis	Original sample (O)	t-value	p-value	Hypothesis support
Compatibility → Introduction intent	0.200	4.615	0.000	Adopted
Competitor pressure → introduction intent	0.320	7.777	0.000	Adopted
Costs → Introduction intent	0.127	3.247	0.001	Rejected
Complexity → Introduction intent	−0.146	4.419	0.000	Rejected
Institutional support → Introduction intent	0.146	3.600	0.000	Adopted
Relative advantage → Introduction intent	0.112	2.233	0.025	Adopted
Security → Introduction intent	−0.127	1.702	0.089	Rejected
Support of management → Introduction intent	0.104	2.781	0.005	Adopted
Task technical suitability → Introduction intent	0.137	3.055	0.002	Adopted

Table 7 Result of model fit analysis

	Standard		Result	
Redundancy	≥0 (Positive number)		Intention to introduce big data	0.155
Model fit (R^2)	0.26~	High	Intention to introduce big data	0.607
	0.13–0.26	Middle		
	0.02–0.13	Low		
Total model fit	0.36~	High	0.668	
	0.25–0.3	Middle		
	0.1–0.25	Low		

introduction is path coefficient of 0.127 and t-value of 3.247. It does not support hypothesis 1 that says it weights negative (−) influence in significant level. The effect of the security on the intention of introduction is path coefficient of 0.127 and t-value of 1.702. It does not support hypothesis 2 that says it weights negative (−) influence in significant level. The influence of relative advantage on the intention of introduction is path coefficient of 0.112 and t-value of 2.333. Hypothesis 3 that says it weighs positive (+) influence in significant level. The effect of the complexity on the intention of introduction is path coefficient of 0.146 and t-value of 4.419. It does not support hypothesis 4 that says it weights positive (+) influence in significant level. The effect of the compatibility on the intention of introduction is path coefficient of 0.200 and t-value of 4.61. Hypothesis 5 that says it weighs positive (+) influence in significant level is adopted.

Second, hypothesis test results between organizational characteristics and intention to introduce big data are analyzed as follows: The impact of the support of

Table 8 Control effect analysis result

Hypothesis	Division	Standardized coefficient	t-value	Hypothesis Verification	Path difference analysis (p-value)	
Compatibility → Introduction intent	1	0.089	1.389	Rejected	0.651	Not different
	2	0.135	2.273	Adopted		
Competitive Pressure → Introduction intent	1	0.350	5.666	Adopted	0.059	Different
	2	0.104	0.767	Rejected		
Cost → Introduction intent	1	0.102	2.217	Adopted	0	Different
	2	−0.258	3.823	Adopted		
Complexity → Introduction intent	1	−0.090	1.772	Adopted	0.201	Not different
	2	0.043	0.398	Rejected		
Institutional Support → Introduction intent	1	0.0713	1.270	Rejected	0.041	Different
	2	0.282	3.027	Adopted		
Relative Advantage → Introduction intent	1	0.161	2.616	Adopted	0.94	Not different

(continued)

Table 8 (continued)

Hypothesis	Division	Standardized coefficient	t-value	Hypothesis Verification	Path difference analysis (p-value)	
	2	0.154	1.977	Adopted		
Security → Introduction intent	1	−0.194	3.775	Adopted	0.028	Different
	2	0.023	0.244	Rejected		
Top Management Support → Introduction intent	1	0.111	1.872	Adopted	0.097	Different
	2	−0.057	0.712	Rejected		
Task Technology Suitability → Introduction intent	1	0.159	2.716	Adopted	0.116	Not different
	2	0.323	3.572	Adopted		

Division: 1 = large/middle-standing companies, 2 = small companies management on the intention of introduction is path coefficient of 0.104 and t-value of 2.781. Hypothesis 6 that says it weighs positive (+) influence in significant level is adopted. The effect of the task technology suitability on the intention of introduction is path coefficient of 0.137 and t-value of 3.055. Hypothesis 7 that says it weighs positive (+) influence in significant level is adopted.

Third, hypothesis test results analysis between environmental characteristics and intention to introduce big data are analyzed as follows: The impact of competitor's pressure on intention to introduce is path coefficient 0.320 and t-value 7.777. Hypothesis 8 that says it weighs positive (+) influence in significant level is adopted. The effect of the institutional support on the intention of introduction is path coefficient of 0.146 and t-value of 3.600. Hypothesis 9 that says it weighs positive (+) influence in significant level is adopted.

Fourth, the analysis results of the hypothesis test of controlling effect of company size is as follows. In the case of competitor pressures, large companies/middle standing companies have a positive impact on the introduction of big data if competitor's pressure is strong, but small companies have insignificant impact. In the case of cost, large companies/middle standing companies have a positive impact on the introduction of big data when prices are high, and small companies respond more sensitively when prices are low. In the case of institutional support, large companies/middle standing companies do not seem to have a significant influence on the introduction of big data, and the governmental institutional support affect big data introduction. In the case of security, large companies/middle standing companies have a positive impact on the introduction of big data when the security is low and there are few restrictions on the use of data, and small companies do not have a big influence on the introduction of big data. In the case of management support, large companies/middle standing companies showed that the active interest and support of CEO influenced big data introduction and did not affect small companies.

5   Conclusions

5.1   Summary of Research and Implications

In order to strengthen the competitiveness of the manufacturing industry, major developed countries are actively promoting smart factory construction. Korea is also striving to build and advance smart factories, but it is still under the basic level. The introduction of big data is very important for building advanced smart factory, but it is not done well and related research is needed. In this study, it is very important to build a smart factory to enhance the competitiveness of manufacturing industry, and analyzed the factors affecting the intention of big data introduction of Korea Industry 4.0 Association members who are leading the fourth industrial revolution in Korea. For the analysis of the research, this study confirmed the empirical analysis targeting the association members who are highly aware of Smart Factory and Big Data.

This study analyzed the impact on companies' intention to introduce big data by deriving companies' awareness of Big Data introduction as technical, organizational, and environmental factors. The independent variables used in this study are the comparison between groups by using technical factors (cost, security, relative advantage, complexity, compatibility), organizational factors (management support, technical suitability), environmental factors (competitor pressure, institutional support), and the company size as control variable, and the dependent variable is intention to introduce big data.

The results of this study are summarized as follows.

First, the relative advantages and compatibility of technical factors had a significant impact on the intention of introducing big data. Cost, security, and complexity have no significant effect on intention to introduce Big Data.

Second, in the organizational factors, the support of the management and the suitability of the task technology were significant in the intention of introducing big data.

Third, in the environmental factors, competitor's pressure and institutional support showed significant results in the intention of introducing big data.

Fourth, variable factors that has the control effect on company size are shown to be competitor's pressure, cost, institutional support, security, management support.

Based on these results, this study is going to suggest the following implications.

First, in the company's introduction of big data, the cost is set to have a negative impact on the intention of introducing big data, but the result of the empirical analysis shows that it has a positive impact. Although security is set to have a negative impact on intention to introduce big data, empirical analysis shows that it has no statistically significant impact. The complexity is set to have a positive impact on the intention of introducing big data, but empirical analysis result showed negative impact. If using big data will make it easier to process work, faster, and qualitatively improved, it will have a positive impact on the introduction of big data. The use of Big Data in the current work will be beneficial from all perspectives, and will have a positive impact on the adoption of Big Data if it helps in the way of work process.

Second, from the organizational viewpoint, if there is the top management support due to necessity of active interest and input resources within the organization, it can positively affect the introduction of big data. If you are able to get the information for the work you need quickly and accurately, it will have a positive impact on big data introduction.

Third, in the environmental aspect, when the advance introduction of competitors leads to falling behind competition and when partner or customer requests the introduction of big data, it will have positive impact on introduction of big data. Company's legal protection for data use and government support for legislation will have a positive impact on the introduction of Big Data.

Fourth, the regulatory effect of the company size is that the pressure of competitors will have a positive impact on the introduction of big data for large/middle standing companies, and little impact on small companies. In the case of cost, large companies/middle standing companies are more likely to trust products with higher prices and have a positive effect on the introduction of big data, and small companies are sensitive to price, which will have a negative impact on the introduction of big data. In the case of institutional support, institutional support of large companies/middle standing companies will be slight in the impact of introduction of big data, but small companies will be under a positive impact on the introduction of big data. In the case of security, large companies/middle standing companies will be under a negative impact on the introduction of big data, if there are many constraints on data utilization, but small companies will have little impact. In the case of interest and support from management, large companies/middle standing companies will have impact on the introduction of big data, but the impact on small companies will be slight.

The practical implications of this study results can be that this study suggested the factors of the companies who concern about building smart factory or consider advancement by empirically verifying them.

5.2 Limitations of Research and Future Research Directions

The limitations of this study are that there are other variables that affect the intention of introducing big data suggested in the model of research. In this study, a research model was established from a big data technical, organizational, and environmental perspective. Theoretical/practical considerations should examine whether there are variables that can explain the relationship of research variables. And, this study failed to include various manufacturing companies by selecting the members of the Korean Industry 4.0 Association as object. As Big Data is still in the development stage and is expected to continue to evolve, positive and negative issues may arise. Future research is needed to expand the scope of smart factory introduction and advancement.

References

1. Kim, H.I.: Smart factory, wings with artificial intelligence. POSRI Issue Report (2017)
2. Park, H.G., Song, H.G., Jang, W.J., Lee, S.R., Lim, C.S.: Fourth Industrial Revolution, Era of New Manufacturing. Heute Books, Seoul (2017)
3. Cho, C.H.: Future of work towards a new social contract. In: The 8th ASIA Future Forum (2017)
4. Jang, W.J., Cho, S.I., Kim, S.S., Gim, G.Y.: A Study on the Implementation of Big Data Infrastructure in Smart Factory. (unpublished)
5. Noh, S.D.: Technology trends and issues of the smart factory Cyber Physical System. J. Korean Inst. Electron. Eng. **43**(6), 481–484 (2016)
6. Larry, P.E.: A Comprehensive Guide to Quality Improvement from the Leading Expert in Information and Data Warehouse Quality, pp. 15–150. Wiley (1999)
7. Svetlana, S.: A framework for evaluating big data initiatives. Gartner. http://www.gartner.com (2014)
8. Raymond, A.H.: Data management regulation: your company needs an up-to-date data/information management policy. Bus. Law Ethics Corner **56**(4), 1–8 (2013)

9. Tankard, C.: Big data security. Netw. Secur. **2012**(7), 5–8 (2012)
10. Zahra, S.A., Hayton, J.C., Neubaum, D.O., Dibrell, C., Craig, J.: Culture of family commitment and strategic flexibility: the moderating effect of stewardship. Entrepreneurship Theory Pract. **32**(6), 1035–1054 (2008)
11. Kim, J.S., Song, T.M.: A study on initial characterization of big data technology acceptance: moderating role of technology user and technology utilizer. J. Korea Contents Assoc. **14**(9), 538–555 (2014)
12. Tornatzky, L.G., Fleischer, M., Chakrabarti, A.K.: The Process of Technological Innovation. Lexington Books, (1990)
13. Go, S.H.: A study on factors motivating cyber security information sharing for responding preemptively to cyber terror threat of national organizations. Doctoral Dissertation, Soongsil University (2016)
14. Han, S.H., Lee, Y.C.: An empirical study on TOE framework based factors for motivation and diffusion of PLM. e-Business Stud. **9**(4), 363–391 (2008)
15. Ka, H.K., Kim, J.S.: A study on the factors affecting the introduction of big data. In: Proceedings of Korea Society of Management Information Systems (2014)
16. Yoon, O.J.: A study on the key factors affecting the diffusion of cybersecurity threat information sharing system. Doctoral Dissertation, Soongsil University (2017)
17. Iacovou, C.L., Benbasat, I., Dexter, A.A.: Electronic data interchange and small organizations: adoption and impact of technology. MIS Q. **19**(4), 465–485 (1995)
18. Lee, S.W.: Research on determinants for big data system adoption in organization. Doctoral dissertation, Sungkyunkwan University (2016)
19. Kuan, K.K.Y., Chau, Y.K.: A perception-based model for EDI adoption in small businesses using a technology-organization-environment framework. Inf. Manag. **38**(8), 507–521 (2001)
20. Yun, K.: The factors affecting the intention to use cloud computing services: Focusing on the financial industry. Doctoral dissertation, Dankook University (2015)
21. Ka, H.K., Kim, J.S.: An empirical study on the influencing factors for big data intented adoption: Focusing on the strategic value recognition and TOE framework. Asia Pac. J. Inf. Syst. **24**(4), 443–472 (2014)
22. Moore, G.C., Benbasat, I.: Development of an instrument to measure the perceptions of adopting an information technology innovation. Inf. Syst. Res. **2**(3), 192–222 (1991)
23. Rogers, E.M.: Diffusion of Innovations, 5th edn. The Free Press, New York (2003)
24. Les, R.: A Summary of Diffusion of Innovations. Enabling Change (2009)
25. Ramamurthy, K., Premkumar, G.: Determinants and outcomes of electronic data interchange diffusion. IEEE Trans. Eng. Manag. **42**(4), 332–351 (1995)
26. Yoon, S.Y.: A study on the factors that influence the intention to use big data from the perspective of the resource based theory. Doctoral dissertation, Dankook University (2016)
27. Lin, A., Chen, N.C.: Cloud computing as an innovation: Perception, attitude, and adoption. Int. J. Inf. Manag. **32**, 533–540 (2012)
28. Lim, J.S.: A study on the effect of the introduction characteristics of cloud computing services on the performance expectancy and the intention to use: Focusing on the innovation diffusion theory. Doctoral dissertation, Dankook University (2012)
29. Premkumar, G., Roberts, M.: Adoption of new information technologies in rural small businesses. Int. J. Manag. Sci. **27**(4), 467–484 (1999)
30. Tornatzky, L.G., Klein, K.J.: Innovation characteristics and innovation adoption-implementation: A meta-analysis of findings. IEEE Trans. Eng. Manag. **29**(1), 28–45 (1982)
31. Abdollahzadehan, A., Hussin, A.R.C., Gohary, M.M., Amini, M.: The organizational critical success factors for adopting cloud computing in SMEs. J. Inf. Syst. Res. Innov. (2013)
32. Alatawi, F.M.H., Dwivedi, Y.K., Williams, M.D., Rana, N.P.: Conceptual Model for Examining Knowledge Management System (KMS) Adoption in Public Sector Organization in Saudi Arabia (2012)
33. Goodhue, D.L.: Understanding user evaluations of information systems. Manag. Sci. **41**(12), 1827–1844 (1995)

34. Lippert, S.K., Govindarajulu, C.: Technological, organization and environment antecedents to the web services adoption. Commun. IIAM **6**(1), 146–158 (2006)
35. Low, C., Chen, Y.: Understanding the determinants of cloud computing adoption. Ind. Manag. Data Syst. **111**(7), 1006–1023 (2011)
36. Zhu, K., Kraemer, K.L., Xu, S.: Electronic business adoption by European firms: a cross-country assessment of the facilitators and inhibitors. Eur. J. Inf. Syst. 251–268 (2003)
37. Oliveira, T., Martins, M.F.: Understanding e-business adoption across industries in European countries. Ind. Manag. Data Syst. **110**(9), 1337–1353 (2010)
38. Zhu, K., Kraemer, K.L., Xu, S.: The process of innovation assimilation by firms in different countries: A technology diffusion perspective on e-business. Manag. Sci. **52**(10), 1557–1576 (2006)
39. Lancaster, S., Yen, D.C., Ku, C.Y.: E-supply Chain management: an evaluation of current web initiatives. Inf. Manag. Comput. Secur. **14**(2), 167–184 (2006)
40. Bakos, J.Y., Michael, E.T.: Information technology and corporate strategy: A research perspective. MIS Q. 107–119 (1986)
41. Shama, A., Citurs, A., Konsynski, B.: Strategic and institutional perspectives in the adoption and early integration of radio frequency identification (RFID). In: Proceeding of the 40th Hawaii International Conference on System Sciences, HICSS 2007. IEEE (2007)
42. Gefen, D., Straub, D.: A practical guide to factorial validity using PLS-graph: Tutorial and annotated example. Commun. Assoc. Inf. Syst. **16**(5), 91–105 (2005)
43. Lee, J.P.: A study on factors influencing intention to use big data in shipping and port companies. Doctoral dissertation, Korea Maritime and Ocean University (2018)
44. Nunnally, J.C., Bernstein, I.H.: Validity. In: Psychometric Theory, pp. 99–132 (1994)
45. Lim, C.S., Han, S.H., Song, H.G., Kim, H.C., Yon, H.S., You, J.Y.: 2016 Smart factory status survey report. Korea Industry 4.0 Innovation & KMAC (2017)
46. Efron, B.: Computers and the theory of statistics: Thinking the unthinkable. SIAM Rev. **21**(4), 460–480 (1979)
47. Tenenhaus, M., Vinzi, V.E., Chatelin, Y.M., Lauro, C.: PLS path modeling. Comput. Stat. Data Anal. **48**(1), 159–205 (2005)
48. Fornell, C., Larcker, D.F.: SEM with unobservable variables and measurement error: Algebra and statistics. J. Mark. Res. **18**(3), 382–388 (1981)

A Study on Factors Affecting the Intension to Use Human Resource Cloud Service

JoongBum Seo, Yong-Won Cho, Kyung-Jin Jung and Gwang-Yong Gim

Abstract This study focuses on the characteristics of Human Resource cloud service and the effects of the intention to use its technology in an empirical manner. The technology's aspects are organized by researching Human Resource Information system and cloud service and Unified Theory of Acceptance and Use of Technology and as guidelines, preceding studies were used to create the research model and propose the hypothesis. This model is based on UTAUT with 8 factors (Compatibility, Ubiquitous Network Access, Complexity, Share ability, Human Resource Information System expertise, Chief Human Resource Officer Support, Competition, Government Policy Support) for performance expectancy, effort expectancy, Facilitating Conditions and Social Influence. For empirical analysis, a survey was conducted on 341 office workers in South Korea. The results are that performance expectancy and effort expectancy were affected by Compatibility, Complexity, Share ability while Ubiquitous Network Access only affected performance expectancy. While facilitating Conditions and Social Influence were affected by Both Chief Human Resource Officer support and Competition, both Human Resource Information System expertise and Government Policy Support didn't affect to facilitating Conditions and Social Influence. Performance expectancy, social influence and facilitating conditions affected the intention to use. This study suggested the practical implications to use human resource cloud service in the future and discussed the limitation of the study and hence forth future research.

Keywords Human resource · HRIS · UTAUT · Cloud · CHRO · HR expertise

J. Seo · Y.-W. Cho · K.-J. Jung · G.-Y. Gim (✉)
Department of Business Administration, Soongsil University, Seoul, South Korea
e-mail: gygim@ssu.ac.kr

J. Seo
e-mail: mathblue9422@ssu.ac.kr

Y.-W. Cho
e-mail: maxcho111@gmail.com

K.-J. Jung
e-mail: seealove@naver.com

1   Introduction

One of the IT technologies that will lead the fourth industry revolution in the 2016 World Economic Forum, cloud services, which enable large-scale data management and processing, shares IT resources and efficiently utilizes idle resources, Moreover, it is to improve work efficiency and productivity through the establishment of an enterprise-wide collaboration system. For this reason, cloud services are becoming a major trend in the IT industry worldwide.

In the past few years, governments have been introducing, planning, and shaping cloud computing with government initiatives. According to a person in charge, there has been boldly investing to take the lead in the cloud ecosystem among global IT companies such as Google, Amazon, Microsoft and IBM.

In spite of this trend, in Korea, there is a perception that it takes a lot of time and money to replace the existing running system with the new one in enterprise information system such as ERP, vague anxiety about security of cloud environment, the lack of laws and regulations on quality and low reliability on quality and performance have been obstacles to cloud adoption.

Therefore, it is hard to find any empirical research on the intention or introduction factor of cloud service in the field of enterprise information system. Moreover, research on cloud service in the field of human resource management is less clear.

In the current enterprise information system, cloud services are still in its infancy. Therefore, it is necessary to study the characteristics of cloud-based enterprise information systems, especially human resource management, and to conduct empirical studies on their intention to use them. In this study, we investigate the characteristics of cloud service and human resource management information system, and try to verify empirically the factors that are important for enterprise to use HR cloud service.

In addition, I would like to suggest some practical implications for the Strategic Usage of HR Cloud Services and the direction of HR Cloud Service.

2   Theoretical Background and Hypotheses

2.1   Cloud Service

In September 2006, Google first proposed cloud computing services as a way for many people to share information resources. Afterwards, major newspapers and journals such as Fortune and InfoWorld and global corporate CEOs pointed to the cloud as their next IT business. In 2009, Gartner defined cloud computing as a form of computing that provides large-scale, unlimited IT-related services to a large number of customers over the Internet [7].

Foster et al. [6] are large-scale distributed computing services provided as a pool of manageable computing, storage, platforms and services that are as abstract, visually

Table 1 Types and characteristics of cloud service tiers

Type	Characteristics
IaaS (Infrastructure as a Service)	Provides IT infrastructure such as storage, server, network, etc. as a service
PaaS (Platform as a Service)	Provide an integrated platform as a service for users to develop and manage necessary software
SaaS (Software as a Service)	Provides services to lease and use various software and applications on various web browsers after purchasing

Table 2 Types and characteristics of cloud service adoption methods

Type	Characteristics
Private cloud	Providing services exclusively within a firewall with strong security features, enhanced security and privacy compared to a public cloud
Public cloud	There is no limit to the use of cloud services. Services with a structure to pay based on usage
Hybrid cloud	The combination of public and private cloud and interoperability

and dynamically expandable as customers demand IT services over the Internet. Cloud services have been defined [6].

Cloud services can generally be categorized by tier or by adoption. However, it is the most common to classify cloud services by hierarchy, and each type and characteristics are shown in Table 1 [17].

In addition, there are many cases in which the introduction methods are also classified. The main types and characteristics are shown in Table 2 [22].

The initial research on cloud services was mainly focused on the theoretical study of the factors that introduced cloud service as a new IT service. Armbrust et al. [3] studied the introduction process, growth process, policy, and work-related content as serious impediments to the transformation of existing information systems into cloud services [3]. Kynetix [12] Service providers tried to consider when making decisions that would not be misleading to service providers [12].

Low and Chen [16] examined eight factors of the TOE framework: complexity, suitability, technology preparation, relative advantage, and top management support to study factors affecting adoption of cloud services for companies in high-tech industries [16]. The results of this study are as follows. First, the effect of the five factors (relative advantage, CEO support, firm size, competitive pressure, and counterparty pressure) on the adoption of cloud services is significant, respectively.

The impact of the intention to use cloud computing service by the public institution is set as the independent variable of security and privacy level, information accessibility, respectively.

In general, among the characteristics of cloud, characteristics of human resource management, we set the cloud characteristics such as suitability, ubiquity, complexity, and sharing as the leading variables of performance expectation and effort expectation and the like.

H1-1: Suitability of personnel cloud service will have a positive effect on performance expectation.
H1-2: Suitability of personnel cloud service will have a positive impact on effort expectations.
H2-1: The ubiquity of HR cloud services will have a positive impact on performance expectations.
H2-2: The ubiquity of HR cloud services will have a positive impact on effort expectations.
H3-1: The complexity of HR cloud services will have a positive impact on performance expectations.
H3-2: The complexity of HR cloud services will have a positive impact on effort expectations.
H4-1: The sharing of the HR cloud service will have a positive impact on performance expectations.
H4-2: The sharing of HR cloud service will have a positive effect on the effort expectation.

2.2 Human Resource Information System

Kavanagh and Johnson [11] is a system used to collect, store, analyze, search and distribute information about the human resources of an organization for human resource management and decision making [11]. It is a Human Resource Information System Lengnick-Hall and Moritz [14] and Gueutal and Stone [9] extended this meaning to electronic Human Resource Management [9, 14].

Until the 1980s, corporations recognized employees as simply objects to manage and developed documents or simple forms of information systems for simple processing such as employee personal information management, attendance record and payroll accounting. However, in recent years, companies have started to recognize employees as strategic resources, and are carrying out talent development, capacity management, talent analysis for employees, and more efficient human resource information system is needed to manage systematically and efficiently.

However, despite the increasing use of human resource information systems, research on the introduction and use of human resource information systems has been performed in a fragmentary manner, and studies related to changes in information technology are insufficient.

In general, human resources information systems are compared with accounting and logistics management systems and we set the characteristics of human resource information system such as HR personnel's expertise Cloud service expertise, the support of HR executives, competitive pressures and government policy support as the leading variables of facilitating conditions and social influences and the like.

H5-1: HR personnel 's expertise Cloud service expertise will have a positive impact on the facilitating conditions of HR cloud services.
H5-2: HR personnel 's expertise Cloud service expertise will have a positive impact on the social impact of HR cloud services.
H6-1: The support of HR executives will have a positive impact on the facilitating conditions of personnel cloud services.
H6-2: Human resources executive support will have a positive impact on the social impact of HR cloud services.
H7-1: Competitive pressures will have a positive impact on the facilitating conditions of HR cloud services.
H7-2: Competitive pressures will have a positive impact on the social impact of HR cloud services.
H8-1: Government policy support will have a positive impact on the facilitating conditions of personnel cloud services.
H8-2: Government policy support will have a positive impact on the social impact of HR cloud services.

2.3   Unified Theory of Acceptance and Use of Technology (UTAUT)

The theory of acceptance of technology is a theory that predicts the attitudes of the users of new information technology and examines the behavioral intentions. Initially, it started with a study of the behavioral intentions and actual behaviors of individuals in social psychology. Individuals or organizations have been applied and developed for the researches related to acceptance of new IT technologies such as cloud, IOT, and AI.

For many years, TAM has been widely used and explored the validity of the new IT technology, but it is not enough to explain the recent acceptance of emerging technologies There are many aspects that have been oversimplified, and there have been many cases where the validity of the research process has not been adequately verified in relation to the external variables added by researchers. In order to overcome the limitations of TAM, Venkatesh et al. [26] proposed the Integrated Technology Acceptance Theory (UTAUT) model, which approaches the acceptance of information technology from the viewpoint of TAM, respectively [26].

The UTAUT model is composed of the TAM-TPB model, the synchronized model (MM), and the PC utilization model (MPCU) integrated with the Rational Behavior Theory (TRA), Planned Behavior Theory (TPB), Technology Acceptance Model (TAM).

UTAUT is a factor that affects the behavioral intention of the use behavior and influences the behavioral intention as the performance expectation, effort expectation, social influence, The relationship between variables is controlled by gender, age, experience, and voluntariness of use [26].

In this paper, we define TAM as a concept that is similar to the perceived usefulness and perceived ease of use of the system, respectively.

Jeon et al. [10] is the result of expectation of performance in public sector's use of cloud computing service [10], Baptista and Oliveira [4] tested the relationship between performance expectation and use intention in the intention of mobile banking acceptance [4] and Kim Jung-seok and Gim [23]. The relationship between performance expectations, effort expectations, and use intentions was tested [23].

Tai and Ku [24] in a study on intention to use mobile stock trading, it was empirically verified that social influence influences intention to use [24]. Lee and Kim [13], demonstrated that external factors such as facilitating conditions and social influences have a significant influence on the intention to use social media in the study of intent to use social media [13].

The results of this study suggest that the performance expectations, effort expectations, facilitating conditions and social influences on the HR cloud service will influence the intention to use.

H9: Performance expectation will have a positive effect on intention to use HR cloud services.
H10: Expectation of effort will have a positive effect on intention to use HR cloud services.
H11: Facilitating conditions will have a positive impact on intention to use HR cloud services.
H10: Social influences will have a positive impact on intention to use HR cloud services.

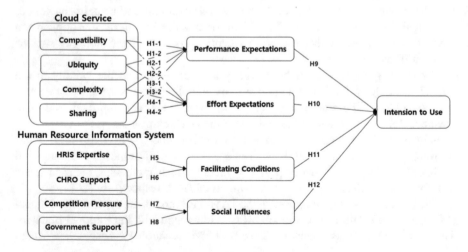

Fig. 1 Research model of human resource cloud service

Based on the above hypotheses, study model in this study has been suggested as shown in Fig. 1.

3 Research Method

3.1 Samples and Data Collection

In order to grasp the intention of using HR cloud service, this research developed a research model based on previous research and set up research hypotheses and conducted empirical analysis to verify each hypothesis.

In order to verify the research model set up in this study, questionnaire data were collected for the employees who work in the companies in Korea. A total of 350 questionnaires were conducted to remove the questionnaire with missing or inappropriate answers, and the final 341 were selected as valid samples.

Table 3 summarizes the sample of this study.

3.2 Measures

In order to measure according to the concept of operational definition of the variables presented in the research model of this study, each item was defined on the basis of previous research.

Table 3 Types and characteristics of cloud service adoption methods

Category and items		Sample size	Ratio (%)
Gender	Male	230	67.4
	Female	141	32.6
Age	Less then 19	1	0.3
	20–29	86	25.2
	30–39	135	39.6
	40–49	117	34.3
	More than 50	2	0.6
Industry	Manufacturing	88	25.8
	Retail	21	6.2
	Service	46	13.5
	Financial	8	2.3
	Public	20	5.9
	Information technology	130	38.1
	The others	28	8.2

Table 4 Source of measuremet items

Dimensions	Construct	Source
Cloud service	Compatibility	[8, 10, 19–21, 25]
	Ubiquity	
	Complexity	
	Trust	
Human resource information system	HR expertise	[1, 15, 18, 20, 25]
	CHRO support	
	Competition	
	Government support	
UTAUT	Performance expectation	[5, 26]
	Effort expectition	
	Facilitating conditions	
	Social influences	
	Usage intension	

First, the preliminary study on the attributes of Cloud Service and Human Resource Information Services are Roger [21], Grover [8], Premkumar and Roberts [20], Teo et al. [25], Jeon and Park [10], Park and Koo [19], Oliveira et al. [18], Lin [15] and [1, 8, 10, 15, 18–21, 25].

And Andersona and Gerbing [2] and Venkatesh et al. [26], which are the preliminary study of the UTAUT model [5, 26].

We measured the effect of performance expectation, effort expectation, facilitating conditions, and social influences on the intention to use.

The sources of this study can be summarized as follows (Table 4).

For each measurement item, a 7-point Likert scale was used.

3.3 Analysis Methods

For the analysis method and measurement tool of structural equation models, this study analyzed the results and verified the hypothesis using SPSS 18.0 and Amos 18.0. For the analysis of the structural equation model, the measurement model was estimated first, and then it was analyzed using the maximum likelihood that is widely used since the two-step approach that estimates the structural model, sample size and the normality assumption were found to be adequate.

4 Analysis and Result

4.1 Measurement Model

This study conducted confirmatory factor analysis to ensure the content validity of the measurement tool. For this, standard, RMSEA, TLI, CFI, and IFI were used to check goodness of fit.

In the initial model, confirmatory factor analysis was performed. The items with high variance of variance with other items due to the fact that the fitness index did not reach the standard level as a whole (personnel cloud service expertise 1, personnel cloud service expertise 3, personnel cloud service expertise 5, Social influence 3, social influence 5, ubiquity 1, ubiquity 5, government policy support 1, government policy support 5, complexity 2, conformity 4, suitability 5, expectation of effort 1, support of HR executives 3, competitive pressure 3, 7, and facilitating conditions 5 [2].

The fit index of the study model after revision index analysis was $\chi 2 = 1572.304$ (p = 0.000), $\chi 2/df = 1.818$, RMSEA = 0.049, RMR = 0.84, GFI = 0.833, AGFI = 0.8, PGFI = 0.696, NFI = 0.96 and IFI = 0.96. GFI was 0.833, but it was relatively low, but all of the other indices were satisfactory, so it was judged that the research model was appropriate as a whole.

Reliability and validity were verified and detailed results are presented in Tables 5 and 6.

4.2 Hypotheses Test

In order to verify each hypothesis of the research model, path analysis was conducted using the AMOS program. Hypothesis 2 that ubiquity among the cloud character-istics variables influenced effort expectations was rejected. And 5-1 and 5-2 were also rejected. Hypothesis 8-1, 8-2 was also dismissed that government policy sup-port affects promotion conditions and social influence. Hypothesis 10 that the effort expectation of the UTAUT variable affects the intention to use was also rejected.

All other hypotheses were adopted, and the results of path analysis using the hypotheses and analysis tools are shown in Table 7.

Table 5 Confirmatory factor analysis base on reliability

Variable	Items of measure	Facto L.D.	Construct reliability	Cromba.Alpha
Compatibility	COMPAT1	0.894	0.934	0.933
	COMPAT2	0.862		
	COMPAT3	0.798		
	COMPAT4	0.521		
	COMPAT 5	0.581		
Ubiquity	UBI1	0.863	0.929	0.934
	UBI2	0.948		
	UBI3	0.914		
	UBI4	0.817		
	UBI5	0.622		
Complexity	COMPLE1	0.881	0.880	0.907
	COMPLE2	0.856		
	COMPLE3	0.793		
	COMPLE4	0.725		
	COMPLE5	0.432		
Sharing	SHA4	0.510	0.892	0.904
	SHA5	0.616		
	SHA6	0.938		
	SHA7	0.827		
HR expertise	HRE1	0.858	0.954	0.967
	HRE2	0.846		
	HRE3	0.846		
	HRE4	0.966		
	HRE5	0.966		
	HRE6	0.916		
CHRO support	CHRO1	0.694	0.947	0.955
	CHRO2	0.874		
	CHRO3	0.727		
	CHRO4	0.636		
	CHRO5	0.694		
Competition	COMPET2	0.646	0.920	0.928
	COMPET3	0.893		
	COMPET4	0.814		
	COMPET5	0.644		

(continued)

Table 5 (continued)

Variable	Items of measure	Facto L.D.	Construct reliability	Cromba.Alpha
Government support	GOV1	0.501	0.952	0.952
	GOV2	0.881		
	GOV3	0.882		
	GOV4	0.935		
	GOV5	0.94		
Performance expectation	PE1	0.75	0.955	0.952
	PE2	0.896		
	PE3	0.832		
	PE4	0.837		
	PE5	0.419		
Effort expectation	EE1	0.406	0.942	0.947
	EE2	0.814		
	EE3	0.653		
	EE4	0.887		
	EE5	0.831		
Facilitating conditions	FAC1	0.477	0.925	0.943
	FAC2	0.49		
	FAC3	0.644		
	FAC4	0.726		
	FAC5	0.733		
Social influences	SOC1	0.917	0.925	0.943
	SOC2	0.876		
	SOC3	0.781		
	SOC4	0.79		
	SOC5	0.669		
Usage intension	USE1	0.565	0.928	0.957
	USE2	0.649		
	USE3	0.605		

Table 6 Discriminant validity

	1	2	3	4	5	6	7	8	9	10	11	12	13
Social influence	0.9												
Compatibility	0.616	0.908											
Ubiquity	0.466	0.483	0.901										
Complexity	0.556	0.625	0.528	0.804									
Sharing	0.534	0.474	0.634	0.654	0.858								
HRIS expertise	0.535	0.48	0.29	0.459	0.27	0.935							
CHRO support	0.746	0.544	0.48	0.554	0.478	0.705	0.904						
Competition	0.614	0.522	0.265	0.515	0.444	0.528	0.556	0.891					
Government support	0.541	0.454	0.163	0.488	0.349	0.54	0.521	0.672	0.932				
Performance expectation	0.701	0.627	0.57	0.624	0.63	0.432	0.664	0.68	0.466	0.899			
Effort expectation	0.707	0.617	0.499	0.627	0.558	0.566	0.68	0.571	0.514	0.747	0.895		
Facilitating conditions	0.842	0.572	0.491	0.571	0.512	0.593	0.813	0.635	0.548	0.711	0.797	0.87	
Usage intension	0.857	0.571	0.522	0.539	0.47	0.568	0.805	0.587	0.486	0.698	0.701	0.815	0.938

[a] AVE (Average Variance Extract)

Table 7 Discriminant validity

Hypothesis	Path coefficient	Standard error	C.R (T)	P values	Result
H1-1	0.349	0.058	6.072	***	Accepted
H1-2	0.367	0.065	5.674	***	Accepted
H2-1	0.151	0.062	2.45	0.014	Accepted
H2-2	0.092	0.069	1.327	0.185	Rejected
H3-1	0.235	0.079	2.962	0.003	Accepted
H3-2	0.373	0.09	4.122	***	Accepted
H4-1	0.315	0.071	4.466	***	Accepted
H4-2	0.2	0.079	2.535	0.011	Accepted
H5-1	−0.061	0.049	−1.242	0.214	Rejected
H5-2	−0.098	0.054	−1.814	0.07	Rejected
H6-1	0.717	0.056	12.694	***	Accepted
H6-2	0.682	0.062	11.082	***	Accepted
H7-1	0.267	0.057	4.691	***	Accepted
H7-2	0.271	0.062	4.341	***	Accepted
H8-1	0.036	0.055	0.66	0.509	Rejected
H8-2	0.071	0.06	1.178	0.239	Rejected
H9	0.127	0.044	2.86	0.004	Accepted
H10	0.044	0.043	1.026	0.305	Rejected
H11	0.294	0.052	5.609	***	Accepted
H12	0.582	0.055	10.63	***	Accepted

5 Conclusions

5.1 Study Results

The purpose of this study is to investigate the characteristics of cloud service and human resource information system that affect the intention of general corporate employees about HR cloud service and to design research model based on previous research related to acceptance of existing technology. Likewise, we analyzed the hypotheses of the research model based on the results of the questionnaires.

The results were as follows.

First, all of the hypotheses that the attributes of the cloud service affect performance expectations were adopted. Recognizing that cloud-based human resource information systems provide more standardized processes and that they can be accessed anytime, anywhere, easily to learn, and can easily share human resource information.

Second, compatibility, complexity, and sharing among the characteristics variables of the cloud service influence the effort expectation. It is easy to use the cloud-based human resource information system to recognize that it is easy to learn and to share the human resources information by providing the standardized HR process provided by the cloud service. But the hypotheses that ubiquity affects effort expectation was rejected.

Third, the support of the HR executives, which is the most influential stakeholder in the human resources information system, affects both facilitating conditions and social influence. The results of this study are as follows: However, the hypothesis that the expertise of the personnel officer on the human resources information system affects the facilitating conditions and the social influence has been rejected. Unlike foreign countries, however, the HR specialist in Korea does not require much expert knowledge on human resource information system.

Forth, the hypothesis that performance expectation, social influence, and facilitating conditions affect intention to use among the variables of UTAUT was adopted, but hypothesis that effort expectation influenced intention to use was rejected. The intention to use the HR cloud service is based on the expectation that the company or organization to which the user belongs will grow in the future and that there will be performance (performance expectation), the evaluation of surroundings of the HR cloud service (social impact) The environmental requirements for use (facilitating conditions) are considered to be affecting factors. However, the effort expectation to HR cloud service is not related to the use of the service. Intention to use is not considered to be an affecting factor in the user's perceived ease of use after introducing the cloud-based HR information system.

This study investigates the characteristics of cloud service and human resource information system which are attracting attention in enterprise information system and suggests the following implications through the empirical study on the relationship between usage intention and use characteristics.

A research model based on Integrated Technology Acceptance Theory (UTAUT) is presented to identify the characteristics that affect the intention to use HR cloud services. Instead of looking at the relationship between UTAUT's independent variables and intention to use, we investigate cloud characteristics that recognize changes when human resource information systems are converted into cloud services and human resource management characteristics of human resource information systems The research model was developed by constructing the hypothesis as the precedent requirement for the performance expectation and the effort expectation by the cloud factor, the promoting condition of the personnel factor, and the factors leading to the social influence.

This research suggests a research model that reflects cloud characteristics and human resource information system characteristics for HR cloud service, and has academic significance as a precedent study in future research on enterprise cloud service, respectively.

5.2 Limitations of Research and Future Directions

This study has conducted an empirical study on cloud - based human resource information system, which is still lacking in academic research. However, there are some limitations.

First, the enterprise cloud service, which is the subject of this study, is still considered to be introduced compared to the personal cloud service. In this study, we conducted an empirical study on the intention to use HR cloud service for the personnel affairs and general users. Since a variety of enterprise information systems with cloud service are widely available, We need empirical studies on the factors affecting the use of actual HR cloud service by designing the model.

Second, the performance characteristics of the UTAUT model used in this study are cloud characteristics, human resource management characteristics for promoting conditions and social influences, and proactive variables, and IT technologies embodied in human resource information system among enterprise cloud services. In this study, we investigate the effect of the factors on the UTAUT variables.

References

1. Allen, D.G., Mahto, R.V., Otondo, R.F.: Web-based recruitment: effects of information, organizational brand, and attitudes towards website on applicant attraction. J. Appl. Psychol. **92**(6), 1696–1708 (2007)
2. Anderson, J.C., Gerbing, D.W.: Structural equation modeling in practice: a review and recommended two-step approach. Psychol. Bull. **103**(3), 411–423 (1988)
3. Armbrust, M., Fox, A., Griffith, R.: A view of cloud computing. Commun. ACM **53**, 50–58 (2010)
4. Baptista, G., Oliveira, T.: Understanding mobile banking: the unified theory of acceptance and use of technology combined with cultural moderators. Comput. Hum. Behav. **50**, 418–430 (2015)
5. Davis, F.D., Bagozzi, R.P., Warshaw, P.R.: User acceptance of computer technology: a comparison of two theoretical models. Manage. Sci. **35**(8), 982–1003 (1989)
6. Foster, I., Zhao, Y., Raicu, I., Lu, S.: Cloud computing and grid computing 360-degree compared. In: Grid Computing Environments Workshop (2008)
7. Gartner: Gartner identifies the top 10 strategic technologies for 2009 (2008)
8. Grover, V.: An empirically derived model for the adoption of customer-based interorganizational systems. Decis. Sci., 603–640 (1993)
9. Gueutal, H.G., Stone, D.L.: The Brave New World of eHR. Jossey-Bass, San Francisco (2005)
10. Jeon, S.H., Park, N.R., Lee, C.C.: Study on the factors affecting the intention to adopt public cloud computing service. Entrue J. Inf. Technol. **10**(2) (2011)
11. Kavanagh, M.J., Thite, M., Johnson, R.D.: Human Resource Information Systems, 3rd edn. Sage, Thousand Oaks (2015)
12. Kynetix Technology Group: Cloud Computing:A strategy guide for board level executive. Kynetix Management Guide (2009)
13. Lee, J.W., Kim, E.H.: Impacts of small and medium enterprises' recognition of social media on their behavioral intention and use behavior. Korea Soc. IT Serv. **14**(1) (2015)
14. Lengnick-Hall, M.L., Moritz, S.: The impact of e-HR on the human resource management function. J. Labor Res. **24**, 365–379 (2003)

15. Lin, H.F.: Understanding the determinants of electronic supply chain management system adoption: using the technology-organization-environment framework. Technol. Forecast. Soc. Chang. **86**, 80–92 (2014)
16. Low, C., Chen, Y.: Understanding the determinants of cloud computing adoption. Ind. Manag. Data Syst. **111**(7), 1006–1023 (2011)
17. Min, O.G., Kim, H.Y., Nam, G.H.: Trends in technology of cloud computing. ETRI **24**(4), 1–13 (2009)
18. Oliveira, T., Thomas, M., Espadanal, M.: Assessing the determinants of cloud computing adoption: an analysis of the manufacturing and services sectors. Inf. Manag. **51**(5), 497–510 (2014)
19. Park, S.C., Koo, C.: A study on end user's intention to use for cloud computing: testing the mediating role of key constructs from UTAUT. J. Int. Electron. Commer. Res. **12**(3) (2012)
20. Premkumar, G., Roberts, M.: Adoption of new information technologies in rural small businesses. Int. J. Manag. Sci. **27**(4), 467–484 (1999)
21. Rogers, E.M.: Diffusion of Innovations, 5th edn. Free Press, New York (2003)
22. Suh, J.-H., Chang, S.-G.: Evaluation of facilitating factors for cloud service by delphi method. Korea Soc. IT Serv. 245–250 (2011)
23. Kim, J., Gim, G.: A study on factors affecting the intention to accept Blockchain technology. Korea Soc. IT Serv. **16**(2) (2017)
24. Tai, Y.M., Ku, Y.C.: Will stock investors use mobile stock trading? A benefit-risk assessment based on a modified UTAUT model. J. Electro. Commer. Res. **14**(1), 67–84 (2013)
25. Teo, T.S.H., Lim, G.S., Fedric, S.A.: The adoption and diffusion of human resources information systems in Singapore. Asia Pac. J. Hum. Resour. **45**(1), 44–62 (2007)
26. Venkatesh, V., Morris, M.G., Davis, G.B., Davis, F.D.: User acceptance of information technology: toward a unified view. MIS Q. **27**(3), 425–478 (2003)

Comparative Analysis of Cost and Elapsed Time of Normalization and De-normalization in the Very Large Database

Seok-Tai Chun, Jihyun Lee and Cheol-Jung Yoo

Abstract Today, data to be processed by information systems is rapidly increasing and complicated, resulting in data integration, standardization, and quality problems. The explosive increase in data is causing performance problems for users seeking the desired information and for operators targeting these users. The industry defines a normalization or de-normalization model and builds a database to solve the performance problems of this very large database. However, it is not well known how they affect actual performance. Therefore, it is necessary to confirm whether the database constructed from the normalized data models and the databases constructed from the data models considering the de-normalization actually contributes to the performance improvement, the development and the simplification of the operation. In this paper, we analyze the effectiveness of de-normalization cost and processing time in the Very Large Database based on the case of establishing database for business to business service of large retailers. As a result, the de-normalized database had 15% faster processing time at a cost of 0.2% of the normalized.

Keywords Database modeling · Database normalization
Database de-normalization · VLDB

1 Introduction

Today, as the field of information is expanded, the data to be processed by the information system is rapidly increasing and complicated, and more than 80% of the software is reported to be related to the database [1]. The Data Base Management

S.-T. Chun · J. Lee (✉) · C.-J. Yoo
Department of Software Engineering, Chonbuk National University, Jeonju, Republic of Korea
e-mail: jihyun30@jbnu.ac.kr

S.-T. Chun
e-mail: stchun@jbnu.ac.kr

C.-J. Yoo
e-mail: cjyoo@kaist.ac.kr

© Springer International Publishing AG, part of Springer Nature 2019
R. Lee (ed.), *Big Data, Cloud Computing, Data Science & Engineering*, Studies
in Computational Intelligence 786, https://doi.org/10.1007/978-3-319-96803-2_13

System (DBMS) did not play a significant role in the days when dozens of gigabytes were large amount of data. It is merely a role of storage that enhances availability and stability, and the programmer has directly processed the data access path (how to input and output data through the DBMS) as logic. Therefore, the focus of the data is on effectively using the program logic rather than organizing it into an effective set to manage. However, currently, systems with several terabytes of data are common, and the performance as well as, technology of the DBMS (parallel processing function, analysis function, distributed processing function, etc.) It is now possible to handle a large amount of data processing that could not be done using tools with a few simple Structured Query Language (SQL) statements.

In large databases, efforts to optimize database performance are essential. In order to optimize database performance, a well-formed database design is required. In the case of a relational DBMS or an object-relational DBMS based on a table, the correct design of the table is most important. Database normalization is a database design technique that simplifies tables by minimizing data redundancy and improves database performance by preventing anomalies when inserting, deleting and updating. It is known that it is practical and practical to design a table as a third normal form (3NF) or a Boyce-Codd normal form (BCNF) to efficiently insert, delete and update data by minimizing data duplication [2].

Normalization is a data modeling technique that focuses on the accuracy and consistency of data by analyzing the dependencies between attributes. As the normalization process progresses from the first normal form to the fifth normal form, the number of tables increases, I/O may be caused, and database performance may be degraded. Therefore, it is argued that the number of tables and disk I/O should be reduced through de-normalization, even if the accuracy and non-redundancy of the data are partially compromised [3]. The de-normalization methods used to reduce the number of tables and the number of disk I/O operations includes performance enhancement through table merging, addition of redundant columns, and table partitioning methods. The purpose of de-normalization is to improve query response time by reducing disk I/O time, so it is often applied in practical work. However, if this de-normalization process depends on the designer's sense it becomes difficult to perform.

Therefore, in designing a database, it is necessary to perform a comparative analysis of performance through practical and objective comparison experiment to decide whether to allow duplication of data through de-normalization or to avoid duplication of data [4]. Therefore, in this paper, we will demonstrate empirically the assumption that a model that effectively applies de-normalization can achieve better performance and simplify its operation than a model with the only normalization through performance analysis of de-normalization model and de-normalization model after de-normalization.

The paper is organized as follows: In Sect. 2, the Online Transaction Processing (OLTP) and Online Analytical Processing (OLAP), normalization and de-normalization are compared. In Sect. 3, we present case studies and data modeling results of empirical experiments. Section 4 describes the experimental results. Finally, Sect. 5 presents the conclusion and future work.

Table 1 Comparison of the OLTP and OLAP

Comparison basis	OLTP	OLAP
Basic	Online transaction system	Online data retrieving and analysis system
Data access	Insert, Update, Delete information from the database	Extract data from the database
Transaction	Short transaction	Long transaction
Query	Simple	Complex
Time	Less	More
Integrity	Must be maintained	Not affected

2 Related Work

2.1 The OLTP and OLAP

In general, the business processing system is called an online transaction system, or OLTP. The OLAP system is a system that analyzes the information already established through OLTP to various points according to the business needs of the company. If the most meaningful element of the system is the data consistency, the OLAP system should be able to show the data to the users effectively in a short period because it uses the information already constructed [5], Table 1 shows the comparison results of the OLTP and OLAP.

2.2 Normalization and De-normalization

The process of structuring data with minimal redundancy in the design of relational databases is called normalization. The goal of database normalization is to reconstruct the ideal relationship to create a small, well-organized relationship. Commonly, normalization involves breaking large, poorly organized tables and relationships into small, well-organized tables and relationships. The goal of normalization is to allow the insertion, deletion, and modification of data in one table to be propagated to the rest of the database due to defined relationships. At first, 1NF [7], 2NF [8] and 3NF normalization were presented, but then a voice-code normalization was proposed [9], and Kent proposed 4NF and 5NF normalization. And the sixth normalization (6NF) was proposed [10].

The database design standard guide should be designed in a way so that the database is fully qualified. Then some may be de-normalized for performance reasons [11]. However, some modeling rules, such as perspective modeling for data warehouse design, recommend exceptionally unqualified designs. In other words, the design on a large scale is not 3NF [12]. De-normalization is the process of attempting to optimize the read performance of a database by adding redundant data or adding result data from group functions [13, 14]. De-normalization techniques are often a means of solving performance and scalability in relational databases for web applications [15]. De-normalization includes table de-normalization and column de-normalization.

In most tasks, the precise data relationship (ensuring the integrity of the data) is an important issue. Therefore, normalization that can guarantee the consistency and integrity of data should be assumed as the basis. However, since it is necessary to consider the complexity of the table and the performance of the system, it is basically aimed to keep the normalized table as it is, and it is necessary to create a view of the problematic table, create a partitioning table, and then apply the clustering method. Then, de-normalization should be considered. If we do indiscriminately de-normalization, it may happen that the integrity of the data is broken, making it impossible to trace or cause inconsistency. For example, if a salesperson's branch location is in a table in a branch office, but is frequently used, it must be changed in the salesperson table as well as in the salesperson table when the branch location is changed. If the point position of the salesperson table is not changed due to an application error or the like and only the point position of the point table is changed, the consistency may be broken. Therefore, it is necessary to select a de-normalized object by selecting a certain criterion. Also, the tables, columns, and relationships that are selected for de-normalization must be managed consistently [5].

3   Use Case Scenario and Data Modeling

In this section, we first introduce a use case scenario used to compare and analyze the performance of normalization and de-normalization models. And we define database requirements based on user scenarios. Based on this requirement, we define the normalized logical/physical model and create the table in the database. Also, the result of defining the de-normalized logical/physical model for the necessary columns and tables in the normalization model is described.

3.1   Use Case Scenario

This use case covers a part of a business to business (B2B) system of a large logistics company. The following scenario is part of the initial requirements definition in the initial stage of the company's distributed information system design.

Our company is a medium distributor of daily necessities that has 148 stores nationwide with its head office in Seoul. The system will be built first for the Seoul head office and the Seoul store, and the system will be extended for the rest of the stores. The headquarters receives the goods from the companies that supply household goods and manages the unit price and inventory. In addition, departments and employees who work in the whole country are managed by the head office, and all staff is responsible for all the work. When a branch receives an order from a customer, it will deliver it to the customer if the product is ordered using a product managed by the head office. In this case, there is a company which is responsible for the delivery, and in our system, only the basic data are managed. Employee information and product information are maintained at the head office, but the same data should be placed in the branch office for quick processing. In order to send DM, address, postal code, telephone number, and e-mail address required for customer information. When mail address is used for mailing, it is necessary to manage various addresses such as home/work/others.

There should be a table that manages only the common code separately. The code used in the table is managed separately from the code management document, and the relationship is not expressed in the ERD. In addition, the code should be able to express values such as 1, 2, and 3, and the depth should be able to express up to 4 levels. Also, it should be possible to check whether or not the node is a terminal node. Orders can be revised, but if the revision occurs, the history should be verifiable, and the order cannot be modified once delivery begins.

3.2 Data Modeling

In this section, ERD is designed based on the user scenario defined in Sect. 3.1. For the main content, conceptual model/normalized logical model/de-normalized logical model is defined individually and finally, table definition, view definition, and index definition is created based on the overall ERD. We use the user scenario of 3.1 to create the conceptual model. Figure 1 defines the common code. Each code is defined by 'GROUP_CODE_ID', and codes corresponding to 'GROUP_CODE_ID' are managed in a common code table as a lower relation. We have created a conceptual model for the business definition by using a COMMON_ CODE_DOMINO table to classify the steps of code.

A data model for storing customer information that can store multiple addresses, zip codes, and phone numbers is defined. The conceptual model of the common code defined in Fig. 1 can cause various problems. It is applied when the domino constraint does not exceed the maximum of 4 steps. If it exceeds 4 steps, an additional property should be considered, and partial modification should be applied in the

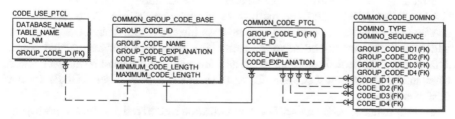

Fig. 1 Normalized data model for common code

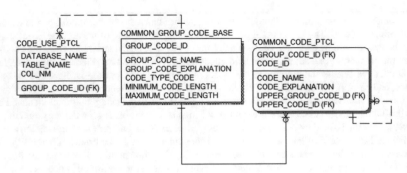

Fig. 2 Normalized data model for customer information

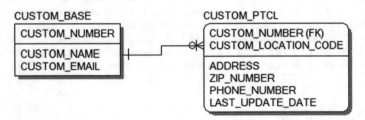

Fig. 3 Normalized data model for customer information

future development process. Therefore, the COMMON_CODE_DOMINO table is deleted, "UPPER GROUP CODE" and "UPPER CODE" are inserted. This allows us to define the parent code and define infinitely many steps.

Figure 2 is a data model for the common code that deletes the COM-MON_CODE DOMINO table in Fig. 1 and adds UPPER_GROUP_CODE_ID and UPPER_CODE_ID to the COMMON_CODE_PTCL table. It resolves the both M:N relationship the constraint problem.

Figure 3 shows the result of normalizing the attributes by separating them into different entity because the address, postal code, and telephone number are repeatedly displayed, and the division values such as home and address cannot be displayed.

Figure 4 shows the de-normalization model for the common code that eliminates the possibilities of performance degradation due to the sub-query of the conditional clause and the scala sub-query of the selection clause by adding the top node.

Figure 5 inserts "Customer Information Location Code (SND_CSTM_LOC_CD)" for mailing to a customer in Fig. 3 so that a DM can be sent only to the representative address instead of sending a DM to all the addresses in the customer description entity. This is the modified normalized result. This column has been de-normalized to have the value of the most recently changed Customer Information Location Code (CUSTOM_LOCATION_CODE) in the customer history.

As a representative example of de-normalization by derivative values, multiplying supply quantity in the order history and product unit price in the product base, the total

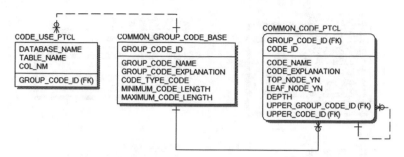

Fig. 4 De-normalized data model for common code

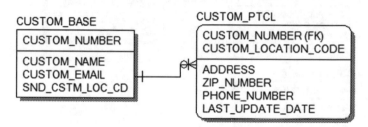

Fig. 5 De-normalized data model for customer information

amount is always available but is frequently accessed items. Frequent joining and reading data can be a major cost and a significant cause of performance degradation. The bigger problem is that when calculating the total amount per order, it is necessary to always combine the order history and the product base to obtain the supply quantity and product unit price per product, and then group them to calculate the complicated calculation. We added the total amount attribute to the order base table, the order history table, and the order history table. Without this de-normalization column, the complexity of development and operation cannot be increased.

Figure 6 has a column titled "TOTAL_PRICE" unlike the order base and order history of the normalized model. To calculate the total amount of orders, we should add the sum of "SUPPLY_QUANTITY" and "PRODUCT_PRICES". The total amount column is a de-normalized column added based on the derived value, and the result is always obtained without performing the joining and calculation between the product details table and the product basic table at the time of order inquiry.

4 Experiments

The total data capacity used in the experiment among the total data capacity of 2.4 TB is about 1 TB. The hardware and software environment for the experiment are shown in Table 2. The execution plans were compared using Quest's Toad for Oracle as the measurement software of the cost and the processing time. Since the trace file

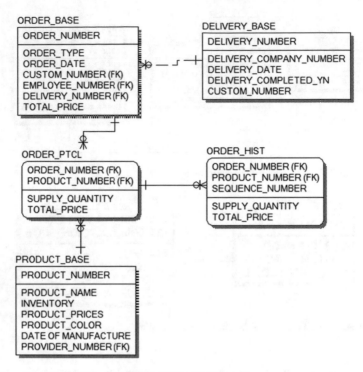

Fig. 6 De-normalized data model added TOTAL_PRICE attribute

Table 2 Hardware and software environment for experiments

Item	Specifications	
H/W	Sever name	HP Z620
	CPU	Intel Xeon E5-2670 (2.8 GHz 8 core 16 thread)
	Memory	32 GB
	Disk1	128 GB SSD
	Raid	LSI 9217-8i (Raid 0)
	Disk2	Western Digital 2 TB * 2
S/W	OS	CentOS 6.7
	DBMS	Oracle 11 g R2 Enterprise Edition
	Tool	Oracle tkprof Quest Toad for Oracle 11.6

analysis is a dump type file, there is a part which is hard to grasp by the human eyes. Those parts were converted into a form by using ORACLE's tkprof.

4.1 Generate Data for Comparative Analysis

We have studied some of the distribution information systems at the time of establishing the B2B system of a large logistics company. It is a general OLTP environment, but the amount of data accumulated every day is large enough to satisfy the requirements of a large-capacity database system. Based on the initial requirements definition, we created a user scenario for this paper. Based on this user scenario, we generated a conceptual model, a normalized logical/physical model, and a de-normalized logical/physical model. The data of the de-normalized model was copied based on this data, and the same data were created by filling the values of the de-normalized column. Based on the data generated, we compiled and analyzed the SQL to produce the same result.

For the performance comparison of normalization and de-normalization, we generated data according to the model created in Sect. 3. The data of the normalized model are first created through the data generation query, then the data of the normalized model is copied, and the value of the de-normalized column is created so that the data of the normalized model and the data of the de-normalized model are kept in the same state respectively. Data generation of the de-normalized physical model was also done through a query.

Case Study 1: Order and customer information management
This Case Study is about the screen that can be seen by the employee, and the main query statement is related to manage order and customer information for each order. This screen is a screen that each employee frequently checks for follow-up action.

Table 3 describes query statements in normalized models and de-normalized models based on the contents defined in Case Study 1. For the comparison/verification, the input value is the employee number and the home address of the customer, and the fixed conditions internally indicate the total amount of sales by order number and the total amount of sales for the orders whose order date is less than the latest one month. This is a query that retrieves the customer's home phone number.

Table 4 is the execution plan based on the query made in Table 3. As depicted in Table 3, the number of tables used for joining in the normalized model is 5, while in the de-normalized model, the number of tables used in joining is reduced to 3. In the de-normalized model, the execution plan was simplified, and the final cost, cardinality, and bytes were also significantly different. The execution plan in the normalized model shows Cost = 3 K, Card = 91, and Bytes = 16 K, while the execution plan in the de-normalized model is less in the de-normalized model with Cost = 322, Card = 1, Bytes = 158, Cardinality, and Bytes.

Case Study 2: Delivery delay management
The company sends an apology to customers once a month with instructions on delivery delays by the postal and e-mail addresses. Instructions are sent to customers who have late shipments last month. The most recently entered or modified mailing address is used.

Table 3 Query for normalization and de-normalization for case study 1

Classification	Query

Normalization	SELECT A.ORD_NO
	    , A.CSTM_NO
	    , E.CSTM_NM
	    , B.PHONE_NO
	    , SUM(C.QTY*D.PRDCT_PRC) AS TOT_PRC
	FROM TN_ORD_BASE A
	    , TN_CSTM_PTCL B
	    , TN_ORD_PTCL C
	    , TN_PRDCT_BASE D
	    , TN_CSTM_BASE E
	WHERE A.CSTM_NO = :IN_CSTM_NO
	AND B.ADDR LIKE :IN_ADDR||'%'
	AND B.CSTM_NO = A.CSTM_NO
	AND B.CSTM_LOC_CD = '1'
	AND A.ORD_DT >= ADD_MONTHS(SYSDATE,-1)
	AND C.ORD_NO = A.ORD_NO
	AND C.PRDCT_NO = D.PRDCT_NO
	AND E.CSTM_NO = B.CSTM_NO
	GROUP BY A.ORD_NO, A.CSTM_NO, E.CSTM_NM, B.PHONE_NO
	ORDER BY A.ORD_NO

De-normalization	SELECT /*+USE_NL(A,C,B) INDEX(C TD_CSTM_BASE_PK)
	INDEX(B TD_CSTM_PTCL_PK)*/
	    A.ORD_NO
	    , A.CSTM_NO
	    , C.CSTM_NM
	    , B.PHONE_NO
	    , A.TOT_PRC
	FROM TD_ORD_BASE A
	    , TD_CSTM_BASE C
	    , TD_CSTM_PTCL B
	WHERE A.EMPL_NO = :IN_EMPL_NO
	AND C.CSTM_NO = A.CSTM_NO
	AND B.ADDR LIKE :IN_ADDR||'%'
	AND B.CSTM_NO = C.CSTM_NO
	AND B.CSTM_LOC_CD = '1'
	AND A.ORD_DT >= ADD_MONTHS(SYSDATE,-1)
	ORDER BY ORD_NO

Based on the contents defined in the Case Study 2, the query statement in the normalized model and the query statement in the de-normalized model were created. The query statement extracts customer number, number of delays in delivery, customer e-mail address, shipping division place, postal code, telephone number, and address. In this normalized model, the number of tables used in the join is five, whereas, in the de-normalized model, the number of tables used in the join is reduced to three. The query in the de-normalized model did not use the subquery to fetch the address, postal code, or phone number of the customer to send the notification, nor did it need to join the order base table to know the delayed customer information.

Table 4 Execution plan for case study 1

Classification	Execution plan

Normalization	0 SELECT STATEMENT Optimizer=ALL_ROWS (Cost=3K Card=91 Bytes=16K)
	  1 0 SORT (GROUP BY) (Cost=3K Card=91 Bytes=16K)
	  2 1 NESTED LOOPS
	  3 2 NESTED LOOPS (Cost=3K Card=91 Bytes=16K)
	  4 3 NESTED LOOPS (Cost=3K Card=91 Bytes=15K)
	  5 4 NESTED LOOPS (Cost=2K Card=17 Bytes=2K)
	  6 5 NESTED LOOPS (Cost=2K Card=17 Bytes=2K)
	  7 6 TABLE ACCESS (BY INDEX ROWID) OF 'TN_ORD_BASE' (TABLE) (Cost=2K Card=16 Bytes=576)
	  8 7 INDEX (RANGE SCAN) OF 'TN_ORD_BASE_I02' (INDEX) (Cost=8 Card=2K)
	  9 6 TABLE ACCESS (BY INDEX ROWID) OF 'TN_CSTM_PTCL' (TABLE) (Cost=1 Card=1 Bytes=69)
	10 9 INDEX (UNIQUE SCAN) OF 'TN_CSTM_PTCL_PK' (INDEX (UNIQUE)) (Cost=0 Card=1)
	11 5 TABLE ACCESS (BY INDEX ROWID) OF 'TD_CSTM_BASE' (TABLE) (Cost=1 Card=1 Bytes=33)
	12 11 INDEX (UNIQUE SCAN) OF 'TD_CSTM_BASE_PK' (INDEX (UNIQUE)) (Cost=0 Card=1)
	13 4 TABLE ACCESS (BY INDEX ROWID) OF 'TN_ORD_PTCL' (TABLE) (Cost=9 Card=5 Bytes=155)
	14 13 INDEX (RANGE SCAN) OF 'TN_ORD_PTCL_PK' (INDEX (UNIQUE)) (Cost=3 Card=5)
	15 3 INDEX (UNIQUE SCAN) OF 'TN_ORD_BASE_PK' (INDEX (UNIQUE)) (Cost=2 Card=1)
	16 2 TABLE ACCESS (BY INDEX ROWID) OF 'TN_PRDCT_BASE' (TABLE) (Cost=3 Card=1 Bytes=16)

De-normalization	0 SELECT STATEMENT Optimizer=ALL_ROWS (Cost=322 Card=1 Bytes=158)
	  1 0 SORT (ORDER BY) (Cost=322 Card=1 Bytes=158)
	  2 1 NESTED LOOPS
	  3 2 NESTED LOOPS (Cost=321 Card=1 Bytes=158)
	  4 3 NESTED LOOPS (Cost=103 Card=1 Bytes=111)
	  5 4 TABLE ACCESS (FULL) OF 'TD_CSTM_PTCL' (TABLE) (Cost=102 Card=1 Bytes=78)
	  6 4 TABLE ACCESS (BY INDEX ROWID) OF 'TD_CSTM_BASE' (TABLE) (Cost=1 Card=1 Bytes=33)
	  7 6 INDEX (UNIQUE SCAN) OF 'TD_CSTM_BASE_PK' (INDEX (UNIQUE)) (Cost=0 Card=1)
	  8 3 BITMAP CONVERSION (TO ROWIDS)
	  9 8 BITMAP AND
	10 9 BITMAP CONVERSION (FROM ROWIDS)
	11 10 INDEX (RANGE SCAN) OF 'TD_ORD_BASE_I01' (INDEX) (Cost=5 Card=2K)
	12 9 BITMAP CONVERSION (FROM ROWIDS)
	13 12 INDEX (RANGE SCAN) OF 'TD_ORD_BASE_I02' (INDEX) (Cost=6 Card=2K)
	14 2 TABLE ACCESS (BY INDEX ROWID) OF 'TD_ORD_BASE' (TABLE) (Cost=321 Card=1 Bytes=47)

```
call      count      cpu    elapsed      disk      query    current       rows
------    ------   ------   --------   -------    -------   --------    -------
Parse          1     0.00       0.00         0          0          0          0
Execute        1     0.05       0.00         0          0          0          0
Fetch          1     0.11       8.95      2363       2681          0         10
------    ------   ------   --------   -------    -------   --------    -------
total          3     0.16       9.00      2363       2681          0         10

Misses in library cache during parse: 1
Optimizer mode: ALL_ROWS
Parsing user id: 91  (JBNU)

Rows     Row Source Operation
------   -------------------------------------------------------------------------
     10  SORT GROUP BY (cr=2681 pr=2263 pw=0 time=0 us cost=2713 size=13651 card=73)
     54   NESTED LOOPS  (cr=2681 pr=2263 pw=0 time=6197608 us)
     54    NESTED LOOPS  (cr=2627 pr=2209 pw=0 time=5527900 us cost=2712 size=13651 card=73)
     54     NESTED LOOPS  (cr=2463 pr=2105 pw=0 time=3514377 us cost=2493 size=12483 card=73)
     10      NESTED LOOPS  (cr=2377 pr=2041 pw=0 time=20610 us cost=2376 size=1820 card=13)
     10       NESTED LOOPS  (cr=2355 pr=2041 pw=0 time=20547 us cost=2363 size=1391 card=13)
     10        TABLE ACCESS BY INDEX ROWID TN_ORD_BASE (cr=2333 pr=2041 pw=0 time=20394 us cost=2350 size=468 card=13)
   2344         INDEX RANGE SCAN TN_ORD_BASE_I02 (cr=8 pr=5 pw=0 time=260 us cost=8 size=0 card=2362)(object id 79978)
     10        TABLE ACCESS BY INDEX ROWID TN_CSTM_PTCL (cr=22 pr=0 pw=0 time=0 us cost=1 size=71 card=1)
     10         INDEX UNIQUE SCAN TN_CSTM_PTCL_PK (cr=12 pr=0 pw=0 time=0 us cost=0 size=0 card=1)(object id 79838)
     10       TABLE ACCESS BY INDEX ROWID TD_CSTM_BASE (cr=22 pr=0 pw=0 time=0 us cost=1 size=33 card=1)
     10        INDEX UNIQUE SCAN TN_CSTM_BASE_PK (cr=12 pr=0 pw=0 time=0 us cost=0 size=0 card=1)(object id 79948)
     54     TABLE ACCESS BY INDEX ROWID TN_ORD_PTCL (cr=86 pr=64 pw=0 time=445055 us cost=9 size=155 card=5)
     54      INDEX RANGE SCAN TN_ORD_PTCL_PK (cr=32 pr=10 pw=0 time=29 us cost=3 size=0 card=5)(object id 79847)
     54    INDEX UNIQUE SCAN TN_PRDCT_BASE_PK (cr=164 pr=104 pw=0 time=0 us cost=2 size=0 card=1)(object id 80646)
     54   TABLE ACCESS BY INDEX ROWID TN_PRDCT_BASE (cr=54 pr=54 pw=0 time=0 us cost=3 size=16 card=1)
```

Fig. 7 Report of executing the query for normalized data model in Case Study 1

Also, the execution plan has been simplified, and the final cost, cardinality, and bytes are also significantly different. The execution plan in the normalized model shows that Cost is 42 K, Card is 6 K, and Byte is 849 K while the execution plan in the de-normalized model shows that 18 K, 1, and 133 respectively.

4.2   Analysis of the Case Study Results

Most of the optimizers run as planned, but occasionally, they run differently from the execution plan of the query. What we need to check is the trace file. The trace file contains the query statement, the execution plan, the execution type actually executed, the CPU utilization, the number of blocks read, and the depth of the data. Because this trace file is inconvenient to read directly, we converted it to a human readable report using the *tkprof* utility. Figure 7 shows the result of executing the query sentence of the normalized model of Case Study 1 and reporting the trace file of execution using the *tkprof* utility.

In the query execution result of the normalized model, the CPU utilization rate is 0.13% (since one CPU is used, it is necessary to multiply the number of CPUs by 16 so that the utilization rate for exactly one CPU can be obtained, that is, 2.08%), the time was 9 s, and the disk used 2,369 blocks. In Fig. 8, we execute the query of the de-normalized model in Case Study 1 and report trace file of execution using the tkprof utility.

In the query result of the de-normalized model, the CPU utilization rate was 0.02% (it is necessary to multiply the number of CPUs by 16 to obtain exactly one CPU utilization rate, that is, actual CPU utilization rate is 0.32%), the time was 0.02 s, and the disk used was 2,209 blocks.

```
call       count      cpu    elapsed      disk      query    current          rows
-------  -------  -------  ---------  --------  ---------  ---------  ------------
Parse          1     0.00       0.00         0          0          0             0
Execute        1     0.00       0.00         0          0          0             0
Fetch          1     0.02       0.02      2209       2383          0            10
-------  -------  -------  ---------  --------  ---------  ---------  ------------
total          3     0.02       0.02      2209       2383          0            10

Misses in library cache during parse: 1
Optimizer mode: ALL_ROWS
Parsing user id: 91  (JBNU)

Rows     Row Source Operation
-------  -----------------------------------------------------------------------
     10  SORT ORDER BY (cr=2383 pr=2209 pw=0 time=1 us cost=2529 size=4983 card=33)
     10   NESTED LOOPS  (cr=2383 pr=2209 pw=0 time=901152 us)
     10    NESTED LOOPS  (cr=2373 pr=2209 pw=0 time=4362813 us cost=2528 size=4983 card=33)
     10     NESTED LOOPS  (cr=2361 pr=2209 pw=0 time=4362705 us cost=2410 size=9440 card=118)
     10      TABLE ACCESS BY INDEX ROWID TD_ORD_BASE (cr=2339 pr=2209 pw=0 time=4362561 us cost=2292 size=5546 card=118)
   2344       INDEX RANGE SCAN TD_ORD_BASE_I02 (cr=13 pr=5 pw=0 time=911 us cost=7 size=0 card=2362)(object id 79986)
     10      TABLE ACCESS BY INDEX ROWID TD_CSTM_BASE (cr=22 pr=0 pw=0 time=0 us cost=1 size=33 card=1)
     10       INDEX UNIQUE SCAN TD_CSTM_BASE_PK (cr=12 pr=0 pw=0 time=0 us cost=0 size=0 card=1)(object id 79948)
     10     INDEX UNIQUE SCAN TD_CSTM_PTCL_PK (cr=12 pr=0 pw=0 time=0 us cost=0 size=0 card=1)(object id 79950)
     10    TABLE ACCESS BY INDEX ROWID TD_CSTM_PTCL (cr=10 pr=0 pw=0 time=0 us cost=1 size=71 card=1)
```

Fig. 8 Report of executing the query for de-normalized data model in Case Study 1

In case of de-normalization through two case studies, meaningful results were obtained. In both tests, 10 experiments were performed to reduce the error range of the experiment. In particular, Case Study 1 uses the employee number that is different every experiment number as an input parameter, so it is not fetched only in the memory but the disk can also read the corresponding block.

Case Study 2 had different results from Case Study 1 in using disk. This is due to the fact that in the process of executing several times in order to reduce the experimental error, Case Study 1 continuously changes the employee number to force the use of the disk. On the other hand, in Case Study 2, since the input value did not have parameters, the result is loaded into the memory by the Least Recently Used algorithm where the experiment is repeated several times. Disk I/O did not happen at all, only fetch because of buffer hit. That is, since it was previously executed and the result of the execution was still in memory, the result of the query was only memory I/O, so only 4,514 records were produced.

In the case of Case Study 2, if the execution plan shows that there are many table access full, and unlike the de-normalized model if we create an execution plan with nested loos join, it results in the increase of recursive call through the sub query clause. We can see that the initial record has been read a lot. In order to produce 4,015 records, table access full of 9,999 records and 29,997 records, respectively, and hash join of two records, respectively, showed a degradation in performance.

Based on the data analyzed in the comparison with the execution plan in Sect. 4.1, it is found that the de-normalized model is more advantageous than the normalized model in query writing and readability, and is more advantageous in the execution plan. Also, based on the data analyzed in the query processing time comparison, it is found that the de-normalized model is more advantageous than the normalized model in CPU utilization, processing time, and disk usage. Table 5 compares the results of the two case studies.

Table 5 Summary of results

Aspects		Normalization	De-normalization
Case 1	CPU usage rate	2.08%	0.32%
	Elapsed time	9 s	0.02 s
	DISK I/O	2369 block	2209 block
	Execution plan	Simple	Complex
	SQL statements	Simple	Complex
Case 2	CPU usage rate	8.96%	1.92%
	Elapsed time	0.61 s	0.08 s
	DISK I/O	0block	0block
	Execution plan	Simple	Complex
	SQL statements	Simple	Complex

5   Conclusion and Future Work

In a small OLTP system environment, performance degradation is not so large even if de-normalization is not performed. In addition, it is more advantageous to take integrity and non-redundancy through a normalized model because the full access table may perform better than the index range scan as the hardware performance of the system increases [6]. However, in a large database, if a full access table occurs, a lot of disk I/O occurs, and the useful data stored in the existing memory can be lost. Therefore, this should be avoided. The explosion of online data has caused many inconveniences and problems for users who want to get the information they want and for businesses that are targeting these users. In order to solve these problems, a number of methodologies have been proposed. Among them, we examined the effectiveness of the de-normalization model in terms of cost and elapsed time.

In this paper, in order to analyze the de-normalization effect of cost and elapsed time in the VLDB, we built a real database based on user scenario and input large amount of data. We then compared and analyzed the results based on normalized and de-normalized models for each case. The results of this study showed that the de-normalized model has less cost and faster elapsed time than the normalized model. Although this experiment was conducted on a distribution system of a large retailer, it did not cover the entire system and did not include the linkage with the delivery system, the logistics system, and the warehouse management system. Therefore, it is necessary to quantitatively and qualitatively analyze and evaluate the effectiveness of de-normalization by extending the comparative studies performed in this experiment to the whole system. Therefore, in future research, we plan to expand the scale of the system and establish a condition that a large number of users can use the system while establishing the whole system such as interworking with other systems and compare and analyze the normalized model and the de-normalized model. In addition, we will analyze the cost and processing time of various queries after adding attributes and tables to the business system to analyze the performance of realistic systems.

Acknowledgements This research was supported by Basic Science Research Program through the National Research Foundation of Korea (NRF) funded by the Ministry of Education (2017R1D1A3B03028609).

References

1. Stephens, R.: Beginning Database Solutions. Wiley Publishing (2009)
2. Verma, S.: Comparing manual and automatic normalization techniques for relation database. Int. J. Res. Eng. Appl. Sci. **2**(1), 59–67 (2012)
3. Lee, J.-S. Lee, C.-H.: Modeling on data performance for very large database. In: Proceedings of Korea Safety Management and Science, pp. 383–391 (2012)
4. Lee, H.-K.: Inefficiency of denormalization in data modeling. Korean Inst. Inf. Sci. Eng. **1**(1), 8–19 (2004)
5. Lee, C.-S.: Database Design and Implementation, Hanbit Publishing Network (2005)
6. Hemalatha, G., Thanuskodi, K.: Optimization of joins using random record generation method. In: Proceedings of the 1st Amrita ACM-W Celebration on Women in Computing in India, Article No. 28 (2010)
7. Codd, E.F.: A relational model of data for large shared data banks. Commun. ACM **13**(6), 377–387 (1970)
8. IBM Research Report RJ909, Republished. In: Rustin, R.J. (ed.) Data base systems: courant computer science symposia series 6. Prentice-Hall (1972)
9. IBM Research Report RJ1385, Republished in Proc. 1974 Congress (Stockholm, Sweden, 1974). North-Holland, N.Y. (1974)
10. Date, C.J., Darwen, H., Lorentzos, N.: Temporal Data and the Relational Model. Morgan Kaufmann (2002)
11. Date, C.J.: Database in Depth: Relational Theory for Practitioners. O'Reilly (2005)
12. Kimball, R.: The Data Warehouse Toolkit, 2nd Ed. Wiley Computer Publishing (2002)
13. Sanders, G.L., Shin, S.K.: Denormalization effects on performance of RDBMS. In: Proceedings of the HICSS Conference, vol. 3 (2001)
14. Shin, S.K., Sanders, G.L.: Denormalization strategies for data retrieval from data warehouses. Decis. Support Syst. **42**(1), 267–282 (2006)
15. Wei, Z., Dejun, J., Pierre, G., Chi, C.-H., van Steen, M.: Service-oriented data denormalization for scalable web applications. In: Proceedings of the International World-Wide Web Conference, pp. 267–276 (2008)

Author Index

© Springer Nature Switzerland AG 2019
R. Lee (ed.), *Big Data, Cloud Computing, Data Science & Engineering*, Studies
in Computational Intelligence 786, https://doi.org/10.1007/978-3-319-96803-2

Printed in the United States
By Bookmasters